Presented to the Library of the Michigan University by The Author.

ELEMENTS

OF

AGRICULTURE

Extract from a letter to the author from Prof. MAPES, editor of the *Working Farmer:*

* * * "After a perusal of your manuscript, I feel authorized in assuring you that, for the use of young farmers, and schools, your book is superior to any other elementary work extant. JAMES J. MAPES."

Letter from the Editor of the N. Y. Tribune:

MY FRIEND WARING,

If all who need the information given in your *Elements of Agriculture* will confess their ignorance as frankly as I do, and seek to dispel it as promptly and heartily, you will have done a vast amount of good by writing it. * * * * * I have found in every chapter important truths, which I, as a would-be-farmer, needed to know, yet which I *did not* know, or had but a confused and glimmering consciousness of, before I read your lucid and straightforward exposition of the bases of Agriculture as a science. I would not have my son grow up as ignorant of these truths as I did for many times the price of your book; and, I believe, a copy of that book in every family in the Union, would speedily add at least ten per cent. per acre to the aggregate product of our soil, beside doing much to stem and reverse the current which now sets so strongly away from the plow and the scythe toward the counter and the office. Trusting that your labors will be widely regarded and appreciated,

I remain yours truly,

HORACE GREELEY.

New York, June 23, 1854.

THE

ELEMENTS OF AGRICULTURE:

A Book for Young Farmers,

WITH QUESTIONS PREPARED FOR THE USE OF SCHOOLS.

BY

GEO. E. WARING, Jr.,

CONSULTING AGRICULTURIST.

The effort to extend the dominion of man over nature is the most healthy and most noble of all ambitions.—BACON.

NEW YORK:
D. APPLETON AND COMPANY,
346 & 348 BROADWAY.
M DCCC LIV.

TO

MY FRIEND AND TUTOR,

PROF. JAMES J. MAPES,

THE PIONEER OF AGRICULTURAL SCIENCE IN AMERICA,

This Book

IS RESPECTFULLY DEDICATED

BY HIS PUPIL,

THE AUTHOR.

TO THE STUDENT.

THIS book is presented to you, not as a work of science, nor as a dry, chemical treatise, but as a plain statement of the more simple operations by which nature produces many results, so common to our observation, that we are thoughtless of their origin. On these results depend the existence of man and the lower animals. No man should be ignorant of their production.

In the early prosecution of the study, you will find, perhaps, nothing to relieve its tediousness; but, when the foundation of agricultural knowledge is laid in your mind so thoroughly that you know the character and use of every stone, then may your thoughts build on it fabrics of such varied construction, and so varied in their uses, that there will be opened to you a new world, even more wonderful and more beautiful than the outward world, which exhibits itself to the senses. Thus may you live two lives, each assisting in the enjoyment of the other.

But you may ask the *practical* use of this. "The world is made up of little things," saith the proverb. So with the productive arts. The steam engine consists of many parts, each part being itself composed of atoms too minute to be detected by our observation. The earth itself, in all its solidity and life, consists entirely of atoms too small to be perceived by the naked eye, each visible particle being an aggregation of thousands of constituent elements. The crop of wheat, which the farmer raises by his labor, and sells for money, is produced by a combination of particles equally small. They are not mysteriously combined, nor irregularly, but each atom is taken from its place of deposit, and carried to its required location in the living plant, by laws as certain as those which regulate the motion of the engine, or the revolutions of the earth.

It is the business of the practical farmer to put together these materials, with the assistance of nature. He may learn her ways, assist her action, and succeed; or he may remain ignorant of her operations, often counteract her beneficial influences, and often fail.

A knowledge of the *inner* world of material things about us will produce pleasure to the thoughtful, and profit to the practical.

CONTENTS.

SECTION FIRST.

THE PLANT.

SECTION SECOND.

THE SOIL.

SECTION THIRD.

MANURES.

SECTION FIRST.

THE PLANT.

SECTION FIRST.

THE PLANT.

CHAPTER I.

INTRODUCTION.

THE object of cultivating the soil is to raise from it a crop of *plants*. In order to cultivate with economy, we must *raise the largest possible quantity with the least expense, and without permanent injury to the soil.*

Before this can be done we must study the character of plants, and learn their exact composition. They are not *created* by a mysterious power, they are merely made up of matters already in existence. They take up water containing food and other mat-

What is the object of cultivating the soil?
What is necessary in order to cultivate with economy?
Are plants created from nothing?

ters, and discharge from their roots those substances that are not required for their growth. It is necessary for us to know what kind of matter is required as food for the plant, and where this is to be obtained, which we can learn only through such means as shall separate the elements of which plants are composed; in other words, we must *take them apart*, and examine the different pieces of which they are formed.

If we burn any vegetable substance it disappears, except a small quantity of earthy matter, which we call *ashes*. In this way we make an important division in the constituents of plants. One portion dissipates into the atmosphere, and the other remains as ashes.

That part which burns away during combustion is called *organic matter*; the ashes are called *inorganic matter*. The organic matter has become air, and hence we conclude that it was originally obtained from air. The inorganic matter has become earth, and was obtained from the soil.

This knowledge can do us no good except by the assistance of chemistry, which explains the properties of each part, and teaches us where it is to be found. It is not necessary for farmers to become chemists. All that is required is, that they should

---

What must we do to learn the composition of plants?
What takes place when vegetable matter is burned?
What do we call the two divisions produced by burning?
Where does organic matter originate? Inorganic?
How much of chemistry should farmers know?

know enough of chemistry to understand the nature of the materials of which their crops are composed, and how those materials are to be used to the best advantage.

This amount of knowledge may be easily acquired, and should be possessed by every person, old or young, whether actually engaged in the cultivation of the soil or not. All are dependent on vegetable productions, not only for food, but for every comfort and convenience of life. It is the object of this book to teach children the first principles of agriculture: and it contains all that is absolutely necessary to an understanding of the practical operations of cultivation, etc.

We will first examine the *organic* part of plants, or that which is driven away during combustion or burning. This matter, though apparently lost, is only changed in form.

It consists of one solid substance, *carbon* (or charcoal), and three gases, *oxygen*, *hydrogen* and *nitrogen*. These four kinds of matter constitute nearly the whole of most plants, the ashes forming often less than one part in one hundred of their dry weight.

When wood is burned in a close vessel, or otherwise protected from the air, its carbon becomes charcoal. All plants contain this substance, it forming

---

Is organic matter lost after combustion?
Of what does it consist?
How large a part of plants is carbon

usually about one half of their dry weight. The remainder of their organic part consists of the three gases named above. By the word gas, we mean *air.* Oxygen, hydrogen and nitrogen, when pure, are always in the form of air. Oxygen has the power of uniting with many substances, forming compounds which are different from either of their constituents alone. Thus: oxygen unites with *iron* and forms oxide of iron or *iron-rust*, which does not resemble the gray metallic iron nor the gas oxygen; oxygen unites with carbon and forms carbonic acid, which is an invisible gas, but not at all like pure oxygen; oxygen combines with hydrogen and forms water. All of the water, ice, steam, etc., are composed of these two gases. We know this because we can artificially decompose, or separate, all water, and obtain as a result simply oxygen and hydrogen, or we can combine these two gases and thus form pure water; oxygen combines with nitrogen and forms nitric acid. These chemical changes and combinations take place only under certain circumstances, which, so far as they affect agriculture, will be considered in the following pages.

As the organic elements of plants are obtained from matters existing in the atmosphere which surrounds our globe, we will examine its constitution.

What do we mean by gas?
Does oxygen unite with other substances?
Give some instances of its combinations

CHAPTER II.

ATMOSPHERE.

Atmospheric air is composed of oxygen and nitrogen. Their proportions are, one part of oxygen to four parts of nitrogen. Oxygen is the active agent in the combustion, decay, and decomposition of organized bodies (those which have possessed animal or vegetable life, that is, organic matter), and others also, in the breathing of animals. Experiments have proved that if the atmosphere consisted of pure oxygen every thing would be speedily destroyed, as the processes of combustion and decay would be greatly accelerated, and animals would be so stimulated that death would soon ensue. The use of the nitrogen in the air is to *dilute* the oxygen, and thus reduce the intensity of its effect.

Besides these two great elements, the atmosphere contains certain impurities which are of great importance to vegetable growth; these are, *carbonic acid, water, ammonia, etc.*

What is atmospheric air composed of?
In what proportions?
What is the use of nitrogen in air?
Does the atmosphere contain other matters useful to vegetation?
What are they?

CARBONIC ACID.

Carbonic acid is in all probability the only source of the carbon of plants, and consequently is of more importance to vegetation than any other single sort of food. It is a gas, and is not, under natural circumstances, perceptible to our senses. It constitutes about $\frac{1}{2500}$ of the atmosphere, and is found in combination with many substances in nature. Marble, limestone and chalk, are carbonate of lime, or carbonic acid and lime in combination ; and carbonate of magnesia is a compound of carbonic acid and magnesia. This gas exists in combination with many other mineral substances, and is contained in all water not recently boiled. Its supply, though small, is sufficient for the purposes of vegetation. It enters the plant in two ways—through the roots in the water which goes to form the sap, and at the leaves, which absorb it from the air in the form of gas. The leaf of the plant seems to have three offices : that of absorbing carbonic acid from the atmosphere—that of assisting in the chemical preparation of the sap—and that of evaporating its water. If we examine leaves with a microscope we shall find that some have as many as 170,000 openings, or

---

What is the source of the carbon of plants?
What is carbonic acid?
What is its proportion in the atmosphere?
Where else is it found?
How does it enter the plant?
What are the offices of leaves?

mouths, in a square inch; others have a much less number. Usually, the pores on the under side of the leaf absorb the carbonic acid. This absorptive power is illustrated when we apply the lower side of a cabbage leaf to a wound, as it draws strongly—the other side of the leaf has no such action. Young sprouts may have the power of absorbing and decomposing carbonic acid.

The roots of plants terminate at their ends in minute spongioles, or mouths for the absorption of fluids containing nutriment. In these fluids there exist greater or less quantities of carbonic acid, and a considerable amount of this gas enters into the circulation of the plant, and is carried to those parts where it is required for decomposition. Plants, under favorable circumstances, may thus obtain about one-third of their carbon.

Carbonic acid, it will be recollected, consists of *carbon and oxygen,* while it supplies only *carbon* to the plant. It is therefore necessary that it be divided, or decomposed, and that the carbon be retained while the oxygen is sent off again into the atmosphere, to reperform its office of uniting with carbon. This decomposition takes place in the *green*

What parts of roots absorb food?
How much of their carbon may plants receive through their roots?
What change does carbonic acid undergo after entering the plant?
In what parts of the plant, and under what influence, is carbonic acid decomposed?

parts of plants and only under the influence of daylight. It is not necessary that the sun shine directly on the leaf or green shoot, but this causes a *more rapid* decomposition of carbonic acid, and consequently we find that plants which are well exposed to the sun's rays make the most rapid growth.

The fact that light is essential to vegetation explains the conditions of different latitudes, which, so far as the assimilation of carbon is concerned, are much the same. At the Equator the days are but about twelve hours long. Still, as the growth of plants is extended over eight or nine months of the year, the duration of daylight is sufficient for the requirements of a luxuriant vegetation. At the Poles, on the contrary, the summer is but two or three months long; here, however, it is daylight all summer, and plants from continual growth develop themselves in that short time.

It will be recollected that carbonic acid constitutes but about $\frac{1}{2500}$ of the air, yet, although about one half of all the vegetable matter in the world is derived from this source, as well as all of the carbon required by the growth of plants, its proportion in the atmosphere is constantly about the same. In order that we may understand this, it becomes necessary for us to consider the means by which it is formed. Carbon, by the aid of fire, is made to

---

Explain the condition of different latitudes.

Does the proportion of carbonic acid in the atmosphere remain about the same?

unite with oxygen, and always when bodies containing carbon are burnt *with the presence of atmospheric air*, the oxygen of that air unites with the carbon, and forms carbonic acid. The same occurs when bodies containing carbon *decay*, as this is simply a slower *burning* and produces the same results. The respiration (or breathing) of animals is simply the union of the carbon of the blood with the oxygen of the air drawn into the lungs, and their breath, when thrown out, always contains carbonic acid. From this we see that the reproduction of this gas is the direct effect of the destruction of all organized bodies, whether by fire, decay, or consumption by animals.

Furnaces are its wholesale manufactories. Every cottage fire is continually producing a new supply, and the blue smoke issuing from the cottage-chimney, as described by so many poets, possesses a new beauty, when we reflect that besides indicating a cheerful fire on the hearth, it contains materials for making food for the cottager's tables and new faggots for his fire. The wick of every burning lamp draws up the carbon of the oil to be made into carbonic acid at the flame. All matters in process of combustion, decay, fermentation, or putrefaction, are returning to the atmosphere those constituents, which they obtained from it. Every living animal, even to the smallest insect, by respiration, spends its life in the

---

Explain some of the operations in which this reproduction takes place.

How is it reproduced?

production of this material necessary to the growth of plants, and at death gives up its body in part for such formation by decay.

Thus we see that there is a continual change from the carbon of plants to air, and from air back to plants, or through them to animals. As each dollar in gold that is received into a country permanently increases its amount of circulating medium, and each dollar sent out permanently decreases it until returned, so the carbonic acid sent into the atmosphere by burning, decay, or respiration, becomes a permanent stock of constantly changeable material, until it shall be locked up for a time, as in a house which may last for centuries, or in an oak tree which may stand for thousands of years. Still, at the decay of either of these, the carbon which they contain must be again resolved into carbonic acid.

The coal-beds of Pennsylvania are mines of carbon once abstracted from the atmosphere by plants. In these coal-beds are often found fern leaves, toads, whole trees, and in short all forms of organized matter. These all existed as living things before the great floods, and at the breaking away of the barriers of the immense lakes, of which our present lakes were merely the deep holes in their beds, they were washed away and deposited in masses so great as to take fire from their chemical changes.

What are the coal-beds of Pennsylvania?
What are often found in them?

It is by many supposed that this fire acting throughout the entire mass (without the presence of air *to supply oxygen* except on the surface) caused it to become melted carbon, and to flow around those bodies which still retained their shapes, changing them to coal without destroying their structures. This coal, so long as it retains its present form, is lost to the vegetable kingdom, and each ton that is burned, by being changed into carbonic acid, adds to the ability of the atmosphere to support an increased amount of vegetation.

Thus we see that, in the provisions of nature, carbon, the grand basis, on which all organized matter is founded, is never permanent in any of its forms. Oxygen is the carrier which enables it to change its condition. For instance, let us suppose that we have a certain quantity of charcoal; this is nearly pure carbon. We ignite it, and it unites with the oxygen of the air, becomes carbonic acid, and floats away into the atmosphere. The wind carries it through a forest, and the leaves of the trees with their millions of mouths drink it in. By the assistance of light it is decomposed, the oxygen is sent off to make more carbonic acid, and the carbon is retained to form a part of the tree. So long as that tree exists in the form of wood, the carbon will re-

---

Explain the manner in which they become coal.
How does the burning of coal benefit vegetation?
Is carbon ever permanent in any of its forms?
What enables it to change its condition?

main unaltered, but when the wood decays, or is burned, it immediately takes the form of carbonic acid, and mingles with the atmosphere ready to be again taken up by plants, and have its carbon deposited in the form of vegetable matter.

The blood of animals contains carbon derived from their food. This unites with the oxygen of the air drawn into the lungs and forms carbonic acid. Without this process, animals could not live. Thus, while by the natural operation of breathing, they make carbonic acid for the uses of the vegetable world, plants, in taking up carbon, throw off oxygen to keep up the life of animals. There is perhaps no way in which we can better illustrate the changes of form in carbon than by describing a simple experiment.

Take a glass tube filled with oxygen gas, and put in it a lump of charcoal, cork the ends of the tube tightly, and pass through the corks the wires of an electrical battery. By passing a stream of electrical fluid over the charcoal it may be ignited, when it will burn with great brilliancy In burning it is dissolved in the oxygen forming carbonic acid, and disappears. It is no more lost, however, than is the carbon of wood which is burned in a stove ; although invisible, it is still in the tube, and may be detected by careful weighing. A more satisfactory proof of its presence may be obtained by *decomposing* the car-

---

Give an instance of such change.
How do plants and animals benefit each other?
Describe the experiment with the glass tube.

bonic acid by drawing the wires a short distance apart, and giving a *spark* of electricity. This immediately separates the oxygen from the carbon which forms a dense black smoke in the tube. By pushing the corks together we may obtain a wafer of charcoal of the same weight as the piece introduced. In this experiment we have changed carbon from its solid form to an invisible gas and back again to a solid, thus fully representing the continual changes of this substance in the destruction of organic matter and the growth of plants.

CHAPTER III.

HYDROGEN, OXYGEN AND NITROGEN.

HYDROGEN AND OXYGEN.

Let us now consider the three gases, *hydrogen*, *oxygen* and *nitrogen*, which constitute the remainder of the organic part of plants.

Hydrogen and oxygen compose *water*, which, if analyzed, yields simply these two gases. Plants perform such analysis, and in this way are able to obtain a sufficient supply of these materials, as their

---

What is water composed of?
If analyzed, what does it yield?
How do plants obtain their hydrogen and oxygen?

sap is composed chiefly of water. Whenever vegetable matter is destroyed by burning, decay, or otherwise, its hydrogen and oxygen unite and form water, which is parted with usually in the form of an invisible vapor. The atmosphere of course contains greater or less quantities of watery vapor arising from this cause and from the evaporation of liquid water. This vapor condenses, forming rains, etc.

Hydrogen and oxygen are never taken into consideration in manuring lands, as they are so readily obtained from the water constituting the sap of the plant, and consequently should not occupy our attention in this book.

NITROGEN.

Nitrogen, the only remaining *organic* constituent of vegetable matter, is for many reasons worthy of close attention.

1. It is necessary to the growth and perfection of all cultivated plants.

2. It is necessary to the formation of animal muscle.

3. It is often deficient in the soil.

4. It is liable to be easily lost from manures.

Although about four fifths of atmospheric air are pure nitrogen, it is almost certain that plants

---

If vegetable matter be destroyed, what becomes of these constituents?

What is the remaining organic constituent?

Why is it worthy of close attention?

Do plants appropriate the nitrogen of the atmosphere?

get no nutriment at all from this source. It is all obtained from some of its compounds, chiefly from the one called ammonia. Nitric acid is also a source from which plants may obtain nitrogen, though to the farmer of less importance than ammonia.

AMMONIA.

Ammonia is composed of nitrogen and hydrogen. It has a pungent smell and is familiarly known as *hartshorn*. The same odor is perceptible around stables and other places where animal matter is decomposing. All animal muscle, certain parts of plants, and other organized substances, consist of compounds containing nitrogen. When these compounds undergo combustion, or are in any manner decomposed, the nitrogen which they contain usually unites with hydrogen, and forms ammonia. In consequence of this the atmosphere always contains more or less of this gas, arising from the decay, etc., which is continually going on all over the world.

This ammonia in the atmosphere is the capital stock to which all plants, not artificially manured, must look for their supply of nitrogen. As they can take up ammonia only through their roots, we must

---

What is the principal source from which they obtain nitrogen?
What is ammonia?
How is it formed?
Where does it always exist?
How do plants take up ammonia?

discover some means by which it may be conveyed from the atmosphere to the soil.

Water may be made to absorb many times its bulk of this gas, and water with which it comes in contact will immediately take it up. Spirits of hartshorn is merely water through which ammonia has been passed until it is saturated.* This power of water has a direct application to agriculture, because the water constituting rains, dews, &c., absorbs the ammonia which the decomposition of nitrogenous matter had sent into the atmosphere, and we find that all rain, snow and dew, contain ammonia. This fact may be chemically proved in various ways, and is perceptible in the common operations of nature. Every person must have noticed that when a summer's shower falls on the plants in a flower garden, they commence their growth with fresh vigor while the blossoms become larger and more richly colored. This effect cannot be produced by watering with spring water, unless it be previously mixed with ammonia, in which case the result will be the same.

Although ammonia is a gas and pervades the atmosphere, few, if any, plants can take it up, as

* By *saturated*, we mean that it contains all that it is capable of holding.

Does water absorb it?
What is *spirits of hartshorn*?
Why is this power of water important in agriculture?
What instance may be cited to prove this?

they do carbonic acid, through their leaves. It must all enter through the roots in solution in the water which goes to form the sap. Although the amount received from the atmosphere is of great importance, there are few cases where artificial applications are not beneficial. The value of farm-yard and other animal manures, depends chiefly on the ammonia which they yield on decomposition. This subject, also the means for retaining in the soil the ammoniacal parts of fertilizing matters, will be fully considered in the section on manures.

After ammonia has entered the plant it may be decomposed, its hydrogen sent off, and its nitrogen retained to answer the purposes of growth. The changes which nitrogen undergoes, from plants to animals, or, by decomposition, to the form of ammonia in the atmosphere, are as varied as those of carbon and the constituents of water. The same little atom of nitrogen may one year form a part of a plant, and the next become a constituent of an animal, or, with the decomposed dead animal, may form a part of the soil. If the animal should fall into the sea he may become food for fishes, and our atom of nitrogen may form a part of a fish. That fish may be eaten by a larger one, or at death may become

---

Can plants use more ammonia than is received from the atmosphere?

On what does the value of animal manure chiefly depend?

What changes take place after ammonia enters the plant?

May the same atom of nitrogen perform many different offices?

food for the whale, through the marine insect, on which it feeds. After the abstraction of the oil from the whale, the nitrogen may, by the putrefaction of his remains, be united to hydrogen, form ammonia, and escape into the atmosphere. From here it may be brought to the soil by rains, and enter into the composition of a plant, from which, could its parts speak as it lies on our table, it could tell us a wonderful tale of travels, and assure us that, after wandering about in all sorts of places, it had returned to us the same little atom of nitrogen which we had owned twenty years before, and which for thousands of years had been continually going through its changes.

The same is true of any of the organic or inorganic constituents of plants. They are performing their natural offices, or are lying in the earth, or floating in the atmosphere, ready to be lent to any of their legitimate uses, sure again to be returned to their starting point.

Thus no atom of matter is ever lost. It may change its place, but it remains for ever as a part of the capital of nature.

Is the same true of the other constituents of plants?
Is any atom of matter ever lost?

CHAPTER IV.

INORGANIC MATTER.

We will now examine the ashes left after burning vegetable substances. This we have called inorganic matter, and it is obtained from the soil. Organic matter, although forming so large a part of the plant, we have seen to consist of four different substances. The inorganic portion, on the contrary, although forming so small a part, consists of no less than *nine* or *ten* different kinds of matter.* These we will consider in order. In their relations to agriculture they may be divided into *three* classes—*alkalies, acids,* and *neutrals.*†

Alkalies and acids are of opposite properties, and when brought together they unite and neutralize each other, forming compounds which are neither alkaline nor acid in their character. Thus, carbonic acid (a gas,) unites with lime—a burning, caustic substance—and forms marble, which is a hard taste-

* Bromine, iodine, etc., are sometimes detected in particular plants, but need not occupy the attention of the farmer.

† This classification is not strictly scientific, but it is one which the learner will find it well to adopt. These bodies are called neutrals because they have no decided alkaline or acid character.

What are ashes called?
How many kinds of matter are there in the ashes of plants?
Into what three classes may they be divided?
What takes place when alkalies and acids are brought together?

less stone. Alkalies and acids are characterized by their desire to unite with each other, and the compounds thus formed have many and various properties, so that the characters of the constituents give no indication of the character of the compound. For instance, lime causes the gases of animal manure to escape, while sulphate of lime (a compound of sulphuric acid and lime) produces an opposite effect, and prevents their escape.

The substances coming under the signification of neutrals, are less affected by the laws of combination, still they often combine feebly with other substances, and some of the resultant compounds are of great importance to agriculture.

ALKALIES.

The alkalies which are found in the ashes of plants are four in number ; they are *potash*, *soda*, *lime* and *magnesia*.

POTASH.

When we pour water over wood ashes it dissolves the *potash* which they contain, and carries it through

---

Is the character of a compound the same as that of its constituents?

Give an instance of this.

Do neutrals combine with other substances?

Name the four alkalies found in the ashes of plants.

in solution. This solution is called *ley*, and if it be boiled to dryness it leaves a solid substance from which pure potash may be made. Potash left exposed to the air absorbs carbonic acid and becomes carbonate of potash, or *pearlash;* if another atom of carbonic acid be added, it becomes super-carbonate of potash, or *salæratus.* Potash has many uses in agriculture.

1. It forms a constituent of nearly all plants.

2. It unites with silica (a neutral), and forms a compound which water can dissolve and carry into the roots of plants; thus supplying them with an ingredient which gives them much of their strength.*

3. It is a strong agent in the decomposition of vegetable matter, and is thus of much importance in preparing manures.

4. It roughens the smooth round particles of sandy soils, and prevents their compacting, as they are often liable to do.

5. It is also of use in killing certain kinds of insects, and, when artificially applied, in smoothing the bark of fruit trees.

The source from which this and the other inor-

* In some soils the *fluorides* uudoubtedly supply plants with soluble silicates, as *fluoric acid* has the power of dissolving silica. Thus, in Derbyshire (England), where the soil is supplied with fluoric acid, grain is said never to lodge.

How may we obtain potash from ashes?
What are some of its agricultural uses?

ganic matters required are to be obtained, will be fully considered in the section on manures.

SODA.

Soda, one of the alkalies contained in the ashes of plants, is very much the same as potash in its agricultural character. Its uses are the same as those of potash—before enumerated. Soda exists very largely in nature, as it forms an important part of common salt, whether in the ocean or in those inland deposits known as rock salt. When combined with sulphuric acid it forms sulphate of soda or *Glauber's salts.* In combination with carbonic acid, as carbonate of soda, it forms the common washing soda of the shops. It is often necessary to render soils fertile.

LIME.

Lime is in many ways important in agriculture:

1. It is a constituent of plants and animals.

2. It assists in the decomposition of vegetable matter in the soil.

3. It corrects the acidity* of sour soils.

* Sourness.

Where is soda found most largely?
What is Glauber's salts?
What is washing soda?
What are some of the uses of lime?

4. As chloride or sulphate of lime it is a good absorbent of fertilizing gases.

In nature it usually exists in the form of carbonate of lime: that is, as marble, limestone, and chalk—these all being of the same composition. In manufacturing caustic (or quick) lime, it is customary to burn the carbonate of lime in a kiln; by this means the carbonic acid is thrown off into the atmosphere and the lime remains in a pure or caustic state. A French chemist states that every cubic yard of limestone that is burned, throws off *ten thousand* cubic yards of carbonic acid, which may be used by plants. This reminds us of the story of Sinbad the sailor, where we read of the immense *genii* who came out of a very small box by the seashore, much to the surprise of Sinbad, who could not believe his eyes, until the *genie* changed himself into a cloud of smoke and went into the box again. Sinbad fastened the lid, and the *genie* must have remained there until the box was destroyed.

Now man is very much like Sinbad, he lets the carbonic acid out from the limestone (when it expands and becomes a gas); and then he raises a crop, the leaves of which drink it in and pack the carbon away in a very small compass as vegetable matter. Here it must remain until the plant is de-

---

How is caustic lime made?
How much carbonic acid is thus liberated?
How does man resemble Sinbad the sailor?

stroyed, when it becomes carbonic acid again, and occupies just as much space as ever.

The burning of limestone is a very prolific source of carbonic acid.

MAGNESIA.

Magnesia is the remaining alkali of vegetable ashes. It is well known as a medicine, both in the form of calcined magnesia, and, when mixed with sulphuric acid, as epsom salts.

Magnesia is necessary to nearly all plants, but too much of it is poisonous, and it should be used with much care, as many soils already contain a sufficient quantity. It is often found in limestone rocks (that class called *dolomites*), and the injurious effects of some kinds of lime, as well as the barrenness of soils made from dolomites, may be attributed entirely to the fact that they contain too much magnesia.

ACIDS.

PHOSPHORIC ACID.

Phosphoric acid.—This subject is one of the greatest interest to the farmer. Phosphoric acid

---

What do you know about magnesia?
What is phosphoric acid composed of?
With what substance does it form its most important compound?

is composed of phosphorus and oxygen. The end of a loco-foco match contains phosphorus, and when it is lighted it unites with the oxygen of the atmosphere and forms phosphoric acid; this constitutes the white smoke which is seen for a moment before the sulphur commences burning. Being an acid, this substance has the power of combining with any of the alkalies. Its most important compound is with lime.

Phosphate of lime forms about 65 per cent. of the dry weight of the bones of all animals, and it is all derived from the soil through the medium of plants. As plants are intended as food for animals, nature has provided that they shall not attain their perfection without taking up a supply of phosphate of lime as well as of the other earthy matters; consequently, there are many soils which will not produce good crops, simply because they are deficient in phosphate of lime. It is one of the most important ingredients of manures, and its value is dependent on certain conditions which will be hereafter explained.

Another use of phosphoric acid in the plant is to supply it with a small amount of *phosphorus*, which seems to be required in the formation of the seed.

Will soils, deficient in phosphate of lime, produce good crops? From what source do plants obtain their phosphorus?

SULPHURIC ACID.

Sulphuric acid is important to vegetation and is often needed to render soils fertile. It is composed of sulphur and oxygen, and is made for manufacturing purposes, by burning sulphur. With lime it forms *sulphate of lime*, which is gypsum or 'plaster.' In this form it is often found in nature, and is generally used in agriculture. Other important methods for supplying sulphuric acid will be described hereafter. It gives to the plant a small portion of *sulphur*, which is necessary to the formation of some of its parts.

NEUTRALS.

SILICA.

This is sand, the base of flint. It is necessary for the growth of all plants, as it gives them much of their strength. In connection with an alkali it constitutes the hard shining surface of corn stalks, straw, etc. Silica unites with the alkalies and forms compounds, such as *silicate of potash, silicate of soda, etc.*, which are soluble in water, and therefore

---

What is sulphuric acid composed of?
What is plaster?
What is silica?
Why is it necessary to the growth of plants?
What compounds does it form with alkalies?

available to plants. If we roughen a corn stalk with sand-paper we may sharpen a knife upon it. This is owing to the hard particles of silica which it contains. Window glass is silicate of potash, rendered insoluble by additions of arsenic and litharge.

Liebig tells us that some persons discovered, between Manheim and Heidelberg in Germany, a mass of melted glass where a hay-stack had been struck by lightning. They supposed it to be a meteor, but chemical analysis showed that it was only the compound of silica and potash which served to strengthen the grass.

There is always *enough* silica in the soil, but it is often necessary to add an alkali to render it available. When grain, etc., lodge or fall down from their own weight, it is altogether probable that they are unable to obtain from the soil a sufficient supply of the soluble silicates, and some form of alkali should be added to the soil to unite with the sand and render it soluble.

CHLORINE.

Chlorine is an important ingredient of vegetable ashes, and is often required to restore the balance to

---

How can you prove its existence in corn stalks?
What instance does Liebig give to show its existence in grass?
How do we supply silicates?
Why does grain lodge?
What is the most important compound of chlorine?

the soil. It is not found alone in nature, but is always in combination with other substances. Its most important compound is with sodium, forming *chloride of sodium* (or common salt). Sodium is the base of soda, and common salt is usually the best source from which to obtain both soda and chlorine. Chlorine unites with lime and forms *chloride of lime*, which is much used to absorb the unpleasant odors of decaying matters, and in this character it is of use in the treatment of manures.

OXIDE OF IRON.

Oxide of iron, one of the constituents of ashes, is common iron rust. *Iron* itself is naturally of a grayish color, but when exposed to the atmosphere, it readily absorbs oxygen and forms a reddish compound. It is in this form that it usually exists in nature, and many soils as well as the red sandstones are colored by it. It is seldom, if ever, necessary to apply this as a manure, there being usually enough of it in the soil.

This red oxide of iron, of which we have been speaking, is called by chemists the *peroxide*. There is another compound which contains less oxygen than

---

Of what use is chloride of lime?
What is oxide of iron?
What is the difference between the *per*oxide and the *prot*oxide of iron?

this, and is called the *protoxide of iron,* which is poisonous to plants. When it exists in the soil it is necessary to use such means of cultivation as shall expose it to the atmosphere and allow it to take up more oxygen and become the peroxide. The black scales which fly from hot iron when struck by the blacksmith's hammer are protoxide of iron.

The *peroxide of iron* is a very good absorbent of ammonia, and consequently, as will be hereafter described, adds to the fertility of the soil.

OXIDE OF MANGANESE, though often found in small quantities in the ashes of cultivated plants, cannot be considered indispensable.

Having now examined all of the materials from which the ashes of plants are formed, we are enabled to classify them in a simple manner, so that they may be recollected. They are as follows:—

ALKALIES.	ACIDS.	NEUTRALS.
Potash.	Sulphuric acid.	Silica.
Soda.	Phosphoric "	Chlorine.
Lime.		Oxide of Iron.
Magnesia.		" Manganese

* There is reason to suppose that *alumina* is an essential constituent of many plants.

What can you say of the oxide of manganese?
How do you classify the inorganic constituents?

CHAPTER V.

GROWTH.

HAVING examined the materials of which plants are made, it becomes necessary to discover how they are put together in the process of growth. Let us therefore suppose a young wheat-plant for instance to be in condition to commence independent growth.

It consists of roots which are located in the soil; leaves which are spread in the air, and a stem which connects the roots and leaves. This stem contains sap vessels (or tubes) which extend from the ends of the roots to the surfaces of the leaves, thus affording a passage for the sap, and consequently allowing the matters taken up to be distributed throughout the plant.

It is necessary that the materials of which plants are made should be supplied in certain proportions, and at the same time. For instance, carbon could not be taken up in large quantities by the leaves, unless the roots, at the same time, were receiving from the soil those mineral matters which are necessary to growth. On the other hand, no con-

---

Of what does a perfect young plant consist?
How must the food of plants be supplied?
Can carbon and earthy matter be taken up at separate stages of growth, or must they both be supplied at once?

siderable amount of earthy matter could be appropriated by the roots unless the leaves were obtaining carbon from the air. This same rule holds true with regard to all of the constituents required; Nature seeming to have made it a law that if one of the important ingredients of the plant is absent, the others, though they may be present in sufficient quantities, cannot be used. Thus, if the soil is deficient in potash, and still has sufficient quantities of all of the other ingredients, the plant cannot take up these ingredients, because potash is necessary to its life.

If a farmer wishes to make a cart he prepares his wood and iron, gets them all in the proper condition, and then can very readily put them together. But if he has all of the *wood* necessary and no *iron*, he cannot make his cart, because bolts, nails and screws are required, and their place cannot be supplied by boards. This serves to illustrate the fact that in raising plants we must give them every thing that they require, or they will not grow at all.

In the case of our young plant the following operations are going on at about the same time.

The leaves are absorbing carbonic acid from the atmosphere, and the roots are drinking in water from the soil.

What seems to be nature's law with regard to this?
What is the similarity between making a cart and raising a crop?
In the growth of a young plant, what operations take place about the same time?

Under the influence of daylight, the carbonic acid is decomposed; its oxygen returned to the atmosphere, and its carbon retained in the plant.

The water taken in by the roots circulates through the sap vessels of the plant, and, from various causes, is drawn up towards the leaves where it is evaporated. This water contains the *nitrogen* and the *inorganic matter* required by the plant and some carbonic acid, while the water itself consists of *hydrogen* and *oxygen.*

Thus we see that the plant obtains its food in the following manner:—

CARBON.—In the form of *carbonic acid* from the atmosphere, and from that contained in the sap, the oxygen being returned to the air.

OXYGEN & HYDROGEN.—From the elements of the water constituting the sap.

NITROGEN.—From the soil (chiefly in form of ammonia). It is carried into the plant through the roots in solution in water.

INORGANIC MATTER.—From the soil, and only *in solution* in water.

What becomes of the carbonic acid?
How is the sap disposed of?
What does it contain?
How does the plant obtain its carbon?
Its oxygen and hydrogen?
Its nitrogen?
Its inorganic matter?

Many of the chemical changes which take place in the interior of the plant are well understood, but they require too much knowledge of chemistry to be easily comprehended by the young learner, and it is not absolutely essential that they should be understood by the scholar who is merely learning the *elements* of the science.

It is sufficient to say that the food taken up by the plant undergoes such changes as are required for its growth; as in animals, where the food taken into the stomach, is digested, and formed into bone, muscle, fat, hair, etc., so in the plant the nutritive portions of the sap are resolved into wood, bark, grain, or some other necessary part.

The results of these changes are of the greatest importance in agriculture, and no person can call himself a *practical farmer* who does not thoroughly understand them.

CHAPTER VI.

PROXIMATE DIVISION OF PLANTS, ETC.

WE have hitherto examined what is called the *ultimate* division of plants. That is, we have looked at each one of the elements separately, and considered its use in vegetable growth.

What changes does the food taken up by the plant undergo?

We will now examine another division of plants, called their *proximate division.* We know that plants consist of various substances, such as wood, gum, starch, oil, etc., and on examination we shall discover that these substances are composed of the various *organic* and *inorganic* ingredients described in the preceding chapters. They are made up almost entirely of *organic* matter, but their ashy parts, though very small, are (as we shall soon see) sometimes of great importance.

These compounds are called *proximate principles,** or *vegetable proximates.* They may be divided into two classes.

The first class are composed of *carbon, hydrogen,* and *oxygen.*

The second class contain the same substances and *nitrogen.*

The first class (those compounds not containing nitrogen) comprise the wood, starch, gum, sugar, and fatty matter which constitute the greater part of all plants, also the acids which are found in sour fruits, etc. Various as are all of these things in their charac-

* By *proximate principle,* we mean that combination of vegetable elements which is known as a vegetable product, such as *wood,* etc.

Of what do wood, starch and the other vegetable compounds chiefly consist?

Are their small ashy parts important?

What are these compounds called?

Into how many classes may proximate principles be divided?

Of what do the first class consist? The second?

What vegetable compounds do the first class comprise?

ters, they are entirely composed of the same ingredients (carbon, hydrogen and oxygen), and usually combined in about the *same proportion.* There may be a slight difference in the composition of their *ashes,* but the organic part is much the same in every case, so much so, that they can often be artificially changed from one to the other.

As an instance of this, it may be recollected by those who attended the Fair of the American Institute, in 1834, that Prof. Mapes exhibited samples of excellent sugar made from the juice of the cornstalk, starch, linen, and woody fibre.

The ease with which these proximates may be changed from one to the other is their most important agricultural feature, and should be clearly understood before proceeding farther. It is one of the fundamental principles on which the growth of both vegetables depends.

The proximates of the first class constitute usually the greater part of all plants, and they are readily formed from the carbonic acid and water which in nature are so plentifully supplied.

The *second class* of proximates, though forming only a small part of the plant, are of the greatest importance to the farmer, being the ones from which

---

Are these substances of about the same composition?
Can they be artificially changed from one to another?
Give an instance of this.
Is the ease with which these changes take place important?
From what may the first class of proximates be formed?

*animal muscle** is made. They consist, as will be recollected, of carbon, hydrogen, oxygen and *nitrogen*, or of *all* of the organic elements of plants. They are all of much the same character, though each kind of plant has its peculiar form of this substance, which is known under the general name of *protein*.

The protein of wheat is called *gluten*—that of Indian corn is *zein*—that of beans and peas is *legumin*. In other plants the protein substances are *vegetable albumen*, *casein*, etc.

Gluten absorbs large quantities of water, which causes it to swell to a great size, and become full of holes. Flour which contains much gluten, makes light, porous bread, and is preferred by bakers, because it absorbs so large an amount of water.

The protein substances are necessary to animal and vegetable life, and none of our cultivated plants will attain maturity (complete their growth), unless allowed the materials required for forming this constituent. To furnish this condition is the object of nitrogen given to plants as manure. If no *nitrogen*

* *Muscle* is *lean meat*, it gives to animals their strength and ability to perform labor.

Why are those of the second class particularly important to farmers?

What is the general name under which they are known?

What is the protein of wheat called?

Why is flour containing much gluten preferred by bakers?

Can protein be formed without nitrogen?

If plants were allowed to complete their growth without a supply of this ingredient, what would be the result?

is supplied the protein substances cannot be formed, and the plant must cease to grow.

When on the contrary *ammonia* is given to the soil (by rains or otherwise), it furnishes nitrogen, while the carbonic acid and water yield the other constituents of protein, and a healthy growth continues, provided that the soil contains the *mineral* matters required in the formation of the ash, in a condition to be useful.

The wisdom of this provision is evident when we recollect that the protein substances are necessary to the formation of muscle in animals, for if plants were allowed to complete their growth without a supply of this ingredient, our grain and hay might not be sufficiently well supplied with it to keep our oxen and horses in working condition, while under the existing law plants must be of nearly a uniform quality (in this respect), and if a field is short of nitrogen, its crop will not be large, and of a very poor quality, but the soil will produce good plants as long as the nitrogen lasts, and then the growth must cease.*

ANIMALS.

That this principle may be clearly understood, it may be well to explain more fully the application of

* This, of course, supposes that the soil is fertile in other respects.

What is the result if a field be deficient in nitrogen?

the proximate constitutents of plants in feeding animals.

Animals are composed (like plants) of organic and inorganic matter, and every thing necessary to build them up exists in plants. It seems to be the office of the vegetable world to prepare the gases in the atmosphere, and the minerals in the earth for the uses of animal life, and to effect this plants put these gases and minerals together in the form of the various *proximates* (or compound substances) which we have just described.

In animals the compounds containing *no nitrogen* comprise the fatty substances, parts of the blood, etc., while the protein compound, or those which *do contain nitrogen*, form the muscle, a part of the bones, the hair, and other portions of the animal.

Animals contain a larger proportion of inorganic matter than plants do. Bones contain a large quantity of phosphate of lime, and we find other inorganic materials performing important offices in the system.

In order that animals may be perfectly developed, they must of course receive as food all of the materials required to form their bodies. They cannot live if fed entirely on one ingredient. Thus, if *starch* alone

---

Of what are the bodies of animals composed?

What is the office of vegetation?

What part of the animal is formed from the first class of proximates? From the second?

Which contains the largest portions of inorganic matter, plants or animals?

Must animals have a variety of food, and why?

be eaten by the animal, he might become *fat*, but his strength would soon fail, because his food contains nothing to keep up the vigor of his *muscles.* If on the contrary the food of an animal consisted entirely of *gluten,* he might be very strong from a superior development of muscle, but would not be fat. Hence we see that in order to keep up the proper proportion of both fat and muscle in our animals (or in ourselves), the food must be such as contains a proper proportion of the two kinds of proximates.

It is for this reason that grain, such as wheat for instance, is so good for food. It contains both classes of proximates, and furnishes material for the formation of both fat and muscle. The value of *flour* depends very much on the manner in which it is manufactured. This will be soon explained.

Apart from the relations between the *proximate principles* of plants, and those of animals, there exists an important relation between their *ashy* or *inorganic* parts; and, food in order to satisfy the demands of animal life, must contain the mineral matter required for the purposes of that life. Take bones for instance. If phosphate of lime is not always supplied in sufficient quantities by food, animals are prevented from the formation of healthy bones. This is particularly

---

Why is grain good for food?

On what does the value of flour depend?

Is there any relation between the ashy part of plants and those of animals?

How may we account for unhealthy bones and teeth?

to be noticed in teeth. Where food is deficient of phosphate of lime, we see poor teeth as a result. Some physicians have supposed that one of the causes of consumption is the deficiency of phosphate of lime in food.

The first class of proximates (starch, sugar, gum, etc.), perform an important office in the animal economy aside from their use in making fat. They constitute the *fuel* which supplies the animal's fire, and gives him his *heat.* The lungs of men and other animals may be called delicate *stoves,* which supply the whole body with heat. But let us explain this matter more fully. If wood, starch, gum, or sugar, be burned in a stove, they produce heat. These substances consist, as will be recollected, of carbon, hydrogen, and oxygen, and when they are destroyed in any way (provided they be exposed to the atmosphere), the hydrogen and oxygen unite and form water, and the carbon unites with the oxygen of the air and forms carbonic acid, as was explained in a preceding chapter. This process is always accompanied by the liberation of *heat,* and the *intensity* of this heat depends on the *time* occupied in its *production.* In the case of decay, the chemical changes take place so slowly that the heat, being conducted away as soon

---

What is a probable cause of consumption?
What is an important use of the first class of proximates?
What may lungs be called?
Explain the production of heat during decomposition.
Why is the heat produced by decay not perceptible?

as formed, is not perceptible to our senses. In combustion (or burning) the same changes take place with much greater rapidity, and the same *amount* of heat being concentrated, or brought out in a far shorter time, it becomes intense, and therefore apparent. In the lungs of animals the same law holds true. The blood contains matters belonging to this carbonaceous class, and they undergo in the lungs the changes which have been described under the head of combustion and decay. Their hydrogen and oxygen unite, and form the moisture of the breath, while their carbon is combined with the oxygen of the air drawn into the lungs, and is thrown out as carbonic acid. The same consequence—heat—results in this, as in the other cases, and this heat is produced with sufficient rapidity for the animal necessities. When an animal exercises violently, his blood circulates with increased rapidity, thus carrying carbon more rapidly to the lungs. The breath also becomes quicker, thus supplying increased quantities of oxygen. In this way the decomposition becomes more rapid, and the animal is heated in proportion.

Thus we see that food has another function besides that of forming animal matter, namely to supply heat. When the food does not contain a sufficient quantity of starch, sugar, etc., to answer

---

Why is the heat produced by combustion apparent?
Explain the production of heat in the lungs of animals?
Why does exercise augment the animal heat?
Under what circumstances is the animal's own fat used in the production of heat?

the demands of the system the *animal's own fat* is carried to the lungs, and there used in the production of heat. This important fact will be referred to again.

CHAPTER VII.

LOCATION OF THE PROXIMATES AND VARIATIONS IN THE ASHES OF PLANTS.

LET us now examine plants with a view to learning the *location* of the various plants.

The stem or trunk of the plant or tree consists almost entirely of *woody fibre* ; this also forms a large portion of the other parts except the seeds, and, in some instances, the roots. The roots of the potato contain large quanties of *starch.* Other roots such as the *carrot* and *turnip* contain *pectic acid,** a nutricious substance resembling starch.

It is in the *seed* however that the more nutritive portions of most plants exist, and here they maintain

* This pectic acid gelatinizes food in the stomach, and thus renders it more digestible.

Of what proximate are plants chiefly composed?
What is the principal constituent of the potato root?
Of the carrot and turnip?
What part of the plant contains usually the most nutriment?

certain relative positions which it is well to understand, and which can be best explained by reference to the following figures, as described by Prof. Johnston:—

Fig. 1.

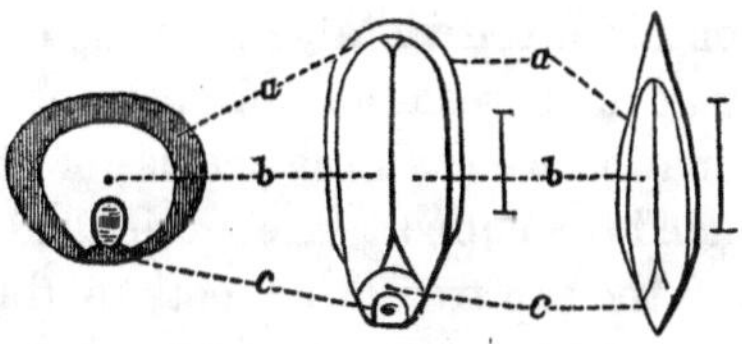

"Thus *a* shows the position of the oil in the outer part of the seed—it exists in minute drops, inclosed in six-sided cells, which consists chiefly of gluten; *b*, the position and comparative quantity of the starch, which in the heart of the seed is mixed with only a small proportion of gluten; *c*, the germ or chit which contains much gluten."*

The location of the *inorganic* part of plants is one of much interest, and shows the adaptation of each part to its particular use. Take a wheat plant, for instance—the stalk, the leaf, and the grain, show in their ashes, important difference of composition. The stalk or straw contains three or four times as large a proportion of ash as the grain, and a no less remarkable difference of composition may be noticed

* See Johnston's Elements, page 41.

Is the composition of the inorganic matter of different parts of the plant the same, or different?

What is the difference between the ash of the straw and that of the grain of wheat?

in the ashes of the two parts. In that of the straw, we find a large proportion of silica and scarcely any phosphoric acid, while in that of the grain there is scarcely a trace of silica, although phosphoric acid constitutes more than one half of the entire weight. The leaves contain a considerable quantity of lime.

This may at first seem an unimportant matter, but on examination we shall see the use of it. The straw is intended to support the grain and leaves, and to convey the sap from the roots to the upper portions of the plant. To perform these offices, *strength* is required, and this is given by the *silica*, and the woody fibre which forms so large a proportion of the stalk. The silica is combined with an alkali, and constitutes the glassy coating of the straw. While the plant is young, this coating is hardly apparent, but as it grows older, as the grain becomes heavier, (verging towards ripeness), the silicious coating of the stalk assumes a more prominent character, and gives to the straw sufficient strength to support the golden head. The straw is not the most important part of the plant as *food*, and therefore requires but little phosphoric acid.

The grain, on the contrary, is especially intended as food, and therefore must contain a large proportion of phosphoric acid—this being, as we have al-

---

What is the reason for this difference?

In what part of the grain does phosphoric acid exist most largely?

ready learned, necessary to the formation of bone—while, as it has no necessity for strength, and as silica is not needed by animals, this ingredient exists in the grain only in a very small proportion. It may be well to observe that the phosphoric acid of grain exists most largely in the hard portions near the shell, or bran. This is one of the reasons why Graham flour is more wholesome than fine flour. It contains all of the nutritive materials which render the grain valuable as food, while flour which is very finely bolted* contains only a small part of the outer portions of the grain (where the phosphoric acid, protein and fatty matters exist most largely). The starchy matter in the interior of the grain, which is the least capable of giving strength to the animal, is carefully separated, and used as food for man, while the better portions, not being ground so finely, are rejected. This one thing alone may be sufficient to account for the fact, that the lives of men have become shorter and less blessed with health and strength, than they were in the good old days when a stone mortar and a coarse sieve made a respectable flour mill.

Another important fact concerning the ashes of plants is the difference of their composition in different plants. Thus, the most prominent ingredient in

* Sifted through a fine cloth called a bolting cloth.

Why is Graham flour more wholesome than fine flour?
Are the ashes of all plants the same in their composition

the ash of the potato is *potash* ; of wheat and other grains, *phosphoric acid* ; of meadow hay, *silica* ; of clover, *lime* ; of beans, *potash*, etc. In grain, *potash* (or *soda*), etc., are among the important ingredients.

These differences are of great importance to the practical farmer, as by understanding what kind of plants use the most of one ingredient, and what kind requires another in large proportion, he can regulate his crops so as to prevent his soil from being exhausted more in one ingredient than in the others, and can also manure his land with reference to the crop which he intends to grow. The tables of analyses in the fifth section will point out these differences accurately.

CHAPTER VIII.

RECAPITULATION.

We have now learned as much about the plant as is required for our immediate uses, and we will carefully reconsider the various points with a view to fixing them permanently in the mind.

Plants are composed of *organic* and *inorganic* matter.

Of what advantage are these differences to the farmer?
Of what are plants composed?

Organic matter is that which burns away in the fire. Inorganic matter is the ash left after burning.

The organic matter of plants consists of three gases, oxygen, hydrogen and nitrogen, and one solid substance carbon (or charcoal). The inorganic matter of plants consists of potash, soda, lime, magnesia, sulphuric acid, phosphoric acid, chlorine, silica, oxide of iron, and oxide of manganese.

Plants obtain their organic food as follows:—Oxygen and hydrogen from water, nitrogen from some compound containing nitrogen (chiefly from ammonia), and carbon from the atmosphere where it exists as carbonic acid—a gas.

They obtain their inorganic food from the soil.

The water which supplies oxygen and hydrogen to plants is readily obtained without the assistance of manures.

Ammonia is obtained from the atmosphere, by being absorbed by rain and carried into the soil, and it enters plants through their roots. It may be artificially supplied in the form of animal manure with profit.

Carbonic acid is absorbed from the atmosphere by leaves, and decomposed in the green parts of plants under the influence of daylight; the carbon is re-

---

What is organic matter? Inorganic?
Of what does organic matter consist? Inorganic?
How do plants obtain their organic food?
How their inorganic?
How is ammonia supplied? Carbonic acid?

tained, and the oxygen is returned to the atmosphere.

When plants are destroyed by decay, or burning, their organic constituents pass away as water, ammonia, carbonic acid, etc., ready again to be taken up by other plants.

The inorganic matters in the soil can enter the plant only when dissolved in water. *Potash, soda, lime,* and *magnesia,* are soluble in their pure forms. Magnesia is injurious when present in too large quantities.

Sulphuric acid is often necessary as a manure, and is usually most available in the form of sulphate of lime or plaster. It is also valuable in its pure form to prevent the escape of ammonia from composts.

Phosphoric acid is highly important, from its frequent deficiency in worn-out soils. It is available only under certain conditions which will be described in the section on manures.

Silica is the base of common sand, and must be united to an alkali before it can be used by the plant, because it is insoluble except when so united.

Chlorine is a constituent of common salt (chloride

When plants are destroyed by combustion or decay, what becomes of their constituents?
How does the inorganic matter enter the plant?
Are the alkalies soluble in their pure forms?
Which one of them is injurious when too largely present?
How may sulphuric acid be supplied?
Is phosphoric acid important?
How must silica be treated?
From what source may we obtain chlorine?

of sodium), and from this source may be obtained in sufficient quantities for manurial purposes.

Oxide of iron is iron rust. There are two oxides of iron, the *peroxide* (red) and the *protoxide* (black). The former is a fertilizer, and the latter poisons plants.

Oxide of manganese is often absent from the ashes of our cultivated plants.

The food of plants, both organic and inorganic, must be supplied in certain proportions, and at the time when it is required. In the plant, this food undergoes such chemical changes as are necessary to growth.

The compounds formed by these chemical combinations are called *proximates.*

Proximates are of two classes, those not containing nitrogen, and those which do contain it.

The first class constitute nearly the whole plant.

The second class, although small in quantity, are of the greatest importance to the farmer, as from them all animal muscle is made.

Animals, like plants, are composed of both organic and inorganic matter, and their bodies are obtained directly or indirectly from plants.

What is the difference between *per*oxide and *proto*xide of iron?
How must the food of plants be supplied?
What takes place after it enters the plant?
What name is given to the compounds thus formed?
How are proximates divided?
Which class constitutes the largest part of the plant?
Of what are animals composed, and how do they obtain the materials from which to form their growth?

The first class of proximates in animals comprise the fat, and like tissues.

The second class form the muscle, hair, gelatine of the bones, etc.

In order that they may be perfectly developed, animals must eat both classes of proximates, and in the proportions required by their natures.

They require the phosphate of lime and other inorganic food which exist in plants.

Seeds are the best adapted to the uses of working animals, because they are rich in all kinds of food required.

Aside from their use in the formation of *fat*, proximates of the first class are employed in the lungs, as fuel to keep up animal heat, which is produced (as in fire and decay) by the decomposition of these substances.

When the food is insufficient for the purposes of heat, the animal's own fat is decomposed, and carried to the lungs as fuel.

The stems, roots, branches, etc., of most plants consist principally of *woody fibre*.

Their seeds, and sometimes their roots, contain considerable quantities of *starch*.

What parts of the animal belong to the first class of proximates?
What to the second?
What is necessary to the perfect development of animals?
Why are seeds valuable for working animals?
What other important use, in animal economy, have proximates of the first class?
Under what circumstances is animal fat decomposed?

The *protein* and the *oils* of most plants exist most largely in the *seeds.*

The location of the proximates, as well as of the inorganic parts of the plant, show a remarkable reference to the purposes of growth, and to the wants of the animal world, as is noticed in the difference between the construction of the straw and that of the kernel of wheat.

The reason why the fine flour now made is not so healthfully nutritious as that which contained more of the coarse portions, is that it is robbed of a large proportion of protein and phosphate of lime, while it contains an undue amount of starch, which is available only to form fat, and to supply fuel to the lungs.

Different plants have ashes of different composition. Thus—one may take from the soil large quantities of potash, another of phosphoric acid, and another of lime.

By understanding these differences, we shall be able so to regulate our rotations, that the soil may not be called on to supply more of one ingredient than of another, and thus it may be kept in balance.

Name the parts of the plant in which the different proximates exist.

State what you know about flour.

Do we know that different plants have ashes of different composition?

The facts contained in this chapter are the *alphabet of agriculture*, and the learner should not only become perfectly familiar with them, but should also clearly understand the *reasons* why they are true, before proceeding further.

How are farmers to be benefited by such knowledge?

SECTION SECOND.

THE SOIL.

SECTION SECOND.

THE SOIL.

CHAPTER I.

FORMATION AND CHARACTER OF THE SOIL.

In the foregoing section, we have studied the character of plants and the laws which govern their growth. We learned that one necessary condition for growth is a fertile soil, and therefore we will examine the nature of different soils, in order that we may understand the relations between them and plants.

The soil is not to be regarded as a mysterious mass of dirt, whereon crops are produced by a mysterious process. Well ascertained scientific

---

What is a necessary condition of growth?

knowledge has proved beyond question that all soils, whether in America or Asia, whether in Maine or California, have certain fixed properties, which render them fertile or barren, and the science of agriculture is able to point out these characteristics in all cases, so that we can ascertain from a scientific investigation what would be the chances for success in cultivating any soil which we examine.

The soil is a great chemical compound, and its chemical character is ascertained (as in the case of plants) by analyzing it, or taking it apart.

We first learn that fertile soils contain both organic and inorganic matter; but, unlike the plant, they usually possess much more of the latter than of the former.

In the plant, the organic matter constitutes the most considerable portion of the whole. In the soil, on the contrary, it usually exists in very small quantities, while the inorganic portions constitute nearly the whole bulk.

The organic part of soils consists of the same materials that constitute the organic part of the plants, and it is in reality decayed vegetable and animal matter. It is not necessary that this organic part of the soil should form any particular proportion

---

What is a fixed character of soils?
How is the chemical character of the soil to be ascertained?
What do we first learn in analyzing a soil?
How do the proportions of organic or inorganic parts of soils compare with those of plants?
Of what does the organic part of soils consist?

of the whole, and indeed we find it varying from one and a half to fifty, and sometimes, in peaty soils, to over seventy per cent. All fertile soils contain some organic matter, although it seems to make but little difference in fertility, whether it be ten or fifty per cent.

The inorganic part of soils is derived from the crumbling of rocks. Some rocks (such as the slates in Central New York) decompose, and crumble rapidly on being exposed to the weather; while granite, marble, and other rocks will last for a long time without perceptible change. The *causes* of this crumbling are various, and are not unimportant to the agriculturist; as by the same processes by which his soil was formed, he can increase its depth, or otherwise improve it. This being the case, we will in a few words explain some of the principal pulverizing agents.

1. The action of frost. When water lodges in the crevices of rocks, and *freezes*, it expands, and bursts the rock, on the same principle as causes it to break a pitcher in winter. This power is very great, and by its assistance, large cannon may be burst. Of course the action of frost is the same on a small scale as when applied to large

---

Can the required proportion be definitely indicated?
From what source is the inorganic part of soils derived?
Do all soils decompose with equal facility?
How does frost affect rocks?
Does it affect soils in the same way?

masses of matter, and, therefore, we find that when water freezes in the *pores** of rocks or stones, it separates their particles and causes them to crumble. The same rule holds true with regard to stiff clay soils. If they are *ridged* in autumn, and left with a rough surface exposed to the frosts of winter, they will become much lighter, and can afterwards be worked with less difficulty.

2. The action of water. Many kinds of rock become so soft on being soaked with water, that they readily crumble.

3. The chemical changes of the constituents of the rock. Many kinds of rock are affected by exposure to the atmosphere, in such a manner, that changes take place in their chemical character, and cause them to fall to pieces. The red kellis of New Jersey (a species of sandstone), is, when first quarried, a very hard stone, but on exposure to the influences of the atmosphere, it becomes so soft that it may be easily crushed between the thumb and finger.

Other actions, of a less simple kind, exert an influence on the stubbornness of rocks, and cause them

* The spaces between the particles.

What is the effect of water on certain rocks?
How are some rocks affected by exposure to the atmosphere?
Give an instance of this.

to be resolved into soils.* Of course, the composition of the soil must be similar to that of the rock from which it was formed; and, consequently, if we know the chemical character of the rock, we can tell whether the soil formed from it can be brought under profitable cultivation. Thus feldspar, on being pulverized, yields potash; talcose slate yields magnesia; marls yield lime, etc.

The soil formed entirely from rock, contains, of course, no organic matter.† Still it is capable of bearing plants of a certain class, and when these die, they are deposited in the soil, and thus form its organic portions, rendering it capable of supporting those plants which furnish food for animals. Thousands of years must have been occupied in preparing the earth for habitation by man.

As the inorganic or mineral part of the soil is usually the largest, we will consider it first.

As we have stated that this portion is formed

* In very many instances the crevices and seams of rocks are permeated by roots, which, by decaying and thus inducing the growth of other roots, cause these crevices to become filled with organic matter. This, by the absorption of moisture, may expand with sufficient power to burst the rock.

† Some rocks contain sulphur, phosphorus, etc., and these may, perhaps, be considered as organic matter.

What is the similarity between the composition of soils and the rocks from which they were formed?
What does feldspar rock yield? Talcose slate? Marls?
Does a soil formed entirely from rock contain organic matter?
How is it affected by the growth of plants?

from rocks, we will examine their character, with a view to showing the different qualities of soils.

As a general rule, it may be stated that *all rocks are either sandstones, limestones, or clays ; or a mixture of two or more of these ingredients.* Hence we find that all mineral soils are either *sandy, calcareous,* (limey), or *clayey ;* or consist of a mixture of these, in which one or another usually predominates. Thus, we speak of a sandy soil, a clay soil, etc. These distinctions (sandy, clayey, loamy, etc.) are important in considering the *mechanical* character of the soil, but have little reference to its fertility.

By *mechanical* character, we mean those qualities which affect the ease of cultivation—excess or deficiency of water, ability to withstand drought, etc. For instance, a heavy clay soil is difficult to plow—retains water after rains, and bakes quite hard during drought ; while a light sandy soil is plowed with ease, often allows water to pass through immediately after rains, and becomes dry and powdery during drought. Notwithstanding those differences in their mechanical character, both soils may be very fertile, or one more so than the other, without reference to the clay and sand which they contain, and which, to *our observation,* form their leading characteristics. The

---

What is the general rule concerning the composition of rocks?
Do these distinctions affect the fertility of soils formed from them?
What do we mean by the mechanical character of the soil?
Is its fertility indicated by its mechanical character?

same facts exist with regard to a loam, a calcareous (or limey) soil, or a vegetable mould. Their mechanical texture is not essentially an index to their fertility, nor to the manures required to enable them to furnish food to plants. It is true, that each kind of soil appears to have some general quality of fertility or barrenness which is well known to practical men, yet this is not founded on the fact that the clay or the sand, or the vegetable matter, enter more largely into the constitution of plants than they do when they are not present in so great quantities, but on certain other facts which will be hereafter explained.

As the following names are used to denote the character of soils, in ordinary agricultural description, we will briefly explain their application:

A *Sandy soil* is, of course, one in which sand largely predominates.

Clay soil, one where *clay* forms a large proportion of the soil.

Loamy soil, where sand and clay are about equally mixed.

Marl contains from five to twenty per cent. of carbonate of lime.

Calcareous soil more than twenty per cent.

Peaty soils, of course, contain large quantities of organic matter.*

* These distinctions are not essential to be learned, but are often convenient.

What is a sandy soil? A clay soil? A loamy soil? A marl? A calcareous soil? A peaty soil?

We will now take under consideration that part of the soil on which depends its ability to supply food to the plant. This portion rarely constitutes more than five or ten per cent. of the entire soil, sometimes less—and it has no reference to the sand, clay, and vegetable matters which they contain. From analyses of many fertile soils, and of others which are barren or of poorer quality, it has been ascertained that the presence of certain ingredients is necessary to fertility. This may be better explained by the assistance of the following table :—

In one hundred pounds.	Soil fertile without manure.	Good wheat soil.	Barren.
Organic matter,	9.7	7.0	4.0
Silica (sand),	64.8	74.3	77.8
Alumina (clay),	5.7	5.5	9.1
Lime,	5.9	1.4	.4
Magnesia,	.9	.7	.1
Oxide of iron,	6.1	4.7	8.1
Oxide of manganese,	.1		.1
Potash,	.2	1.7	
Soda,	.4	.7	
Chlorine,	.2	.1	
Sulphuric acid,	.2	.1	
Phosphoric acid,	.4	.1½	
Carbonic acid,	4.0		
Loss during the analysis	1.4	3.6½	.4
	100.0	100.0	100.0

How large a part of the soil may be used as food by plants?
What do we learn from the analyses of barren and fertile soils?

The soil represented in the first column might still be fertile with less organic matter, or with a larger proportion of clay (alumina), and less sand (silica). These affect its *mechanical* character; but, if we look down the column, we notice that there are small quantities of lime, magnesia, and the other constituents of the ashes of plants (except ox. of manganese). It is not necessary that they should be present in the soil in the exact quantity named above, but *not one must be entirely absent, or greatly reduced in proportion.* By referring to the third column, we see that these ingredients are not all present, and the soil is barren. Even if it were supplied with all but one or two, potash and soda for instance, it could not support a crop without the assistance of manures containing these alkalies. The reason for this must be readily seen, as we have learned that no plant can arrive at maturity without the necessary supply of materials required in the formation of the ash, and these materials can be obtained only from the soil; consequently, when they do not exist there, it must be barren.

The inorganic part of soils has two distinct offices to perform. The clay and sand form a

---

What can you say of the soils represented in the table of analyses?

What proportion of the fertilizing ingredients is required?

If the soil represented in the third column contained all the ingredients required except potash and soda, would it be fertile?

What would be necessary to make it so?

What is the reason for this?

What are the offices performed by the inorganic part of soils?

mass of material into which roots can penetrate, and thus plants are supported in their position. These parts also absorb heat, air and moisture to serve the purposes of growth, as we shall see in a future chapter. The minute portions of soil, which comprise the acids, alkalies, and neutrals, furnish plants with their ashes, and are the most necessary to the fertility of the soil.

GEOLOGY.

The relation between the inorganic part of soils and the rocks from which it was formed, is the the foundation of Agricultural Geology. Geology may be briefly named the *science of rocks.* It would not be proper in an elementary work to introduce much of this study, and we will therefore simply state that the same kind of rock is of the same composition all over the world; consequently, if we find a soil in New England formed from any particular rock, and a soil from the same rock in Asia, their natural fertility will be the same in both localities. Some rocks consist of a mixture of different kinds of minerals; and some, consisting chiefly of one ingredient, are of different degrees of *hardness.* Both of these changes must affect the character of the soil, but it may be laid down as rule that, *when the rocks of two loca-*

What is geology?
Is the same kind of rock always of the same composition?
How do rocks differ?

tions are exactly alike, the soils formed from them will be of the same natural fertility, and in proportion as the character of rocks changes, in the same proportion will the soils differ.

In most districts the soil is formed from the rock on which it lies; but this is not always the case. Soils are often formed by deposits of matter brought by water from other localities. Thus the alluvial banks of rivers consist of matters brought from the country through which the rivers have passed. The river Nile, in Egypt, yearly overflows its banks, and deposits large quantities of mud brought from the uninhabited upper countries. The prairies of the West owe a portion of their soil to deposits by water. Swamps often receive the washings of adjacent hills; and, in these cases, their soil is derived from a foreign source.

We might continue to enumerate instances of the relations between soils and the sources whence they originated, thus demonstrating more fully the importance of geology to the farmer; but it would be beyond the scope of this work, and should be investigated by scholars more advanced than those who are studying merely the *elements* of agricultural science.

The mind, in its early application to any branch

What rule may be given in relation to soils formed from the same or different rocks?
Are all soils formed from the rocks on which they lie?
What instances can you give of this?

of study, should not be charged with intricate subjects. It should master well the *rudiments,* before investigating those matters which should *follow* such understanding.

By pursuing the proper course, it is easy to learn all that is necessary to form a good foundation for a thorough acquaintance with the subject. If this foundation is laid thoroughly, the learner will regard plants and soils as old acquaintances, with whose formation and properties he is as familiar as with the construction of a building or simple machine. A simple spear of grass will become an object of interest, forming itself into a perfect plant, with full development of roots, stem, leaves, and seeds, by processes with which he feels acquainted. The soil will cease to be mere dirt; it will be viewed as a compound substance, whose composition is a matter of interest, and whose care is productive of intellectual pleasure. The commencement of study in any science must necessarily be wearisome to the young mind, but its more advanced stages amply repay the trouble of early exertions.

In what light will plants and soils be regarded by those who understand them?

CHAPTER II.

USES OF ORGANIC MATTER.

It will be recollected that, in addition to its mineral portions, the soil contains organic matter in varied quantities. It may be fertile with but one and a half per cent. of organic matter, and some peaty soils contain more than fifty per cent. or more than one half of the whole.

The precise amount necessary cannot be fixed at any particular sum; perhaps five parts in a hundred would be as good a quantity as could be recommended.

The soil obtains its organic matter in two ways. First, by the decay of roots and dead plants, also of leaves, which have been brought to it by wind, etc. Second, by the application of organic manures.

When a crop of clover is raised, it obtains its carbon from the atmosphere; and, if it be plowed under, and allowed to decay, a portion of this carbon is deposited in the soil. Carbon constitutes nearly the whole of the dry weight of the clover, aside from the constituents of water; and, when we calculate the immense quantity of hay, and roots grown on

---

What proportion of organic matter is required for fertility?
How does the soil obtain its organic matter?
How does the growth of clover, etc., affect the soil?

an acre of soil in a single season, we shall find that the amount of carbon thus deposited is immense. If the clover had been removed, and the roots only left to decay, the amount of carbon deposited would still have been very great. The same is true in all cases where the crop is removed, and the roots remain to form the organic or vegetable part of the soil. While undergoing decomposition, a portion of this matter escapes in the form of gas, and the remainder chiefly assumes the form of carbon (or charcoal), in which form it will always remain, without loss, unless driven out by fire. If a bushel of charcoal be mixed with the soil now, it will be the same bushel of charcoal, neither more nor less, a thousand years hence, unless some influence is brought to bear on it aside from the growth of plants. It is true that, in the case of the decomposition of organic matter in the soil, certain compounds are formed, known under the general names of *humus* and *humic acid*, which may, in a slight degree, affect the growth of plants, but their practical importance is of too doubtful a character to justify us in considering them. The application of manures, containing organic matter, such as peat, muck, animal manure, etc., supplies the soil with carbon on the same principle, and the decomposing matters also

---

When organic matter decays in the soil, what becomes of it?
Is charcoal taken up by plants?
Are humus and humic acid of great practical importance?

generate * carbonic acid gas while being decomposed. The agricultural value of carbon in the soil depends (as we have stated), not on the fact that it enters into the composition of plants, but on certain other important offices which it performs, as follows :—

1. It makes the soil more retentive of manures.
2. It causes it to appropriate larger quantities of the fertilizing gases of the atmosphere.
3. It gives it greater power to absorb moisture.
4. It renders it warmer.

1. Carbon (or charcoal) makes the soil retentive of manures, because it has in itself a strong power to absorb, and retain † fertilizing matters. There is a simple experiment by which this power can be shown.

Ex.—Take two barrels of pure beach sand, and mix with the sand in one barrel a few handfuls of charcoal dust, leaving that in the other pure. Pour the brown liquor of the barn-yard through the pure sand, and it will pass out at the bottom unaltered. Pour the same liquor through the barrel, containing the charcoal, and pure water will be obtained as a result. The reason for this is that the

* Produce.

† By absorbing and retaining, we mean taking up and holding.

On what does the agricultural value of the carbon in the soil depend?

Why does it make the soil more retentive of manure?

What is the experiment with the barrels of sand?

charcoal retains all of the impurities of the liquor, and allows only the water to pass through. Charcoal is often employed to purify water for drinking, or for manufacturing purposes.

A rich garden-soil contains large quantities of carbonaceous matter ; and, if we bury in such a soil a piece of tainted meat or a fishy duck, it will, in a short time, be deprived of its odor, because the charcoal in the soil will entirely absorb it.

Carbon absorbs gases as well as the impurities of water ; and, if a little charcoal be sprinkled over manure, or any other substance, emitting offensive odors, the gases escaping will be taken up by the charcoal, and the odor will cease.

It has also the power of absorbing *mineral* matters, which are contained in water. If a quantity of salt water be filtered through charcoal, the salt will be retained, and the water will pass through pure.

We are now able to see how carbon renders the soil retentive of manures.

1st. Manures, which resemble the brown liquor of barn-yards, have their fertilizing matters taken out, and retained by it.

Will charcoal purify water?

If a piece of tainted meat, or a fishy duck be buried in a rich garden soil, what takes place?

What is the reason of this?

How does charcoal overcome offensive odors?

How can you prove that charcoal absorbs the *mineral* impurities of water?

2d. The gases arising from the decomposition (*rotting*) of manure are absorbed by it.

3d. The soluble mineral portions of manure, which might in some soils leach down with water, are arrested and retained at a point at which they can be made use of by the roots of plants.

2. Charcoal in the soil causes it to appropriate larger quantities of the fertilizing gases of the atmosphere, on account of its power, as just named, to absorb gases.

The atmosphere contains results, which have been produced by the breathing of animals and by the decomposition of various kinds of organic matter, which are exposed to atmospheric influences. These gases are chiefly ammonia and carbonic acid, both of which are largely absorbed by water, and consequently are contained in rain, snow, etc., which, as they enter the soil, give up these gases to the charcoal, and they there remain until required by plants. Even the air itself, in circulating through the soil, gives up fertilizing gases to the carbon, which it may contain.

3. Charcoal gives to the soil power to absorb moisture, because it is itself one of the best ab-

---

How does charcoal in the soil affect the manures applied?
Why does charcoal in the soil cause it to appropriate the gases of the atmosphere?
What fertilizing gases exist in the atmosphere?
How are they carried to the soil?
Does the carbon retain them after they reach the soil?
What can you say of the air circulating through the soil?
How does carbon give the soil power to absorb moisture?

sorbents in nature; and it has been proved by accurate experiment that peaty soils absorb moisture with greater rapidity, and part with it more slowly than any other kind.

4. Carbon in the soil renders it warmer, because it darkens its color. Black surfaces absorb more heat than light ones, and a black coat, when worn in the sun, is warmer than one of a lighter color. By mixing carbon with the soil, we darken its color, and render it capable of absorbing a greater amount of heat from the sun's rays.

It will be recollected that, when vegetable matter decomposes in the soil, it produces certain gases (carbonic acid, etc.), which either escape into the atmosphere, or are retained in the soil for the use of plants. The production of these gases is always accompanied by *heat,* which, though scarcely perceptible to our senses, is perfectly so to the growing plant, and is of much practical importance. This will be examined more fully in speaking of manures.

Another important part of the organic matter in the soil is that which contains *nitrogen.* This forms but a very small portion of the soil, but it is of the greatest importance to vegetables. As the nitrogen in food is of absolute necessity to the growth of

---

How does it render it warmer?

Is the heat produced by the decomposition of organic matter perceptible to our senses?

Is it so to the growing plant?

What is another important part of the organic matter in the soil?

animals, so the nitrogen in the soil is indispensable to the growth of cultivated plants. It is obtained by the soil in the form of ammonia (or nitric acid), from the atmosphere, or by the application of animal matter. In some cases, manures called *nitrates* * are used ; and, in this manner, nitrogen is given to the soil.

We have now learned that the organic matter in the soil performs the following offices :—

Organic matter thoroughly decomposed is *carbon*, and has the various effects ascribed to this substance on p. 79.

Organic matter in process of decay produces carbonic acid, and sometimes ammonia in the soil ; also its decay causes heat.

Organic matter containing *nitrogen*, such as animal substances, etc., furnish ammonia, and other nitrogenous substances to the roots of plants.

* Nitrates are compounds of nitric acid (which consists of nitrogen and oxygen), and alkaline substances. Thus nitrate of potash (saltpetre), is composed of nitric acid and potash: nitrate of soda (cubical nitre), of nitric acid and soda.

How is it obtained by the soil?
What offices does the organic matter in the soil perform?

CHAPTER III.

USES OF INORGANIC MATTER.

THE offices performed by the inorganic constituents of the soil are many and important.

These, as well as the different conditions in which the bodies exist, are necessary to be thoroughly studied.

Those parts which constitute the larger proportion of the soil, namely the clay, sand, and limy portions, are useful for purposes which have been named in the first part of this section, while the *clay* has an additional effect in the absorption of ammonia.

For this purpose, it is as effectual as charcoal, the gases escaping from manures, as well as those existing in the atmosphere, and in rain-water, being arrested by clay as well as charcoal.*

The more minute ingredients of the soil—those which enter into the construction of plants—exist in conditions which are more or less favorable or in-

* It is due to our country, as well as to Prof. Mapes and others, who long ago explained this absorptive power of clay and carbon, to say that the subject was perfectly understood and practically applied in America a number of years before Prof. Way published the discovery in England as original.

What effect has clay besides the one already named?
How does it compare with charcoal for this purpose?

jurious to vegetable growth. The principal condition necessary to fertility is *capacity to be dissolved,* it being (so far as we have been able to ascertain) a fixed rule, as was stated in the first section, that *no mineral substance can enter into the roots of a plant except it be dissolved in water.*

The *alkalies* potash, soda, lime, and magnesia, are in nearly all of their combinations in the soil sufficiently soluble for the purposes of growth.

The *acids* are, as will be recollected, sulphuric and phosphoric. These exist in the soil in combination with the alkalies, as sulphates and phosphates, which are more or less soluble under natural circumstances. Phosphoric acid in combination with lime as phosphate of lime is but slightly soluble; but, when it exists in the compound known as *super*-phosphate of lime, it is much more soluble, and consequently enters into the composition of plants with much greater facility. This matter will be more fully explained in the section on manures.

The *neutrals*, silica, chlorine, oxide of iron, and oxide of manganese, deserve a careful examination. Silica exists in the soil usually in the form of *sand,* in which it is, as is well known, perfectly insoluble; and, before it can be used by plants, which often re-

---

What particular condition of inorganic matter is requisite for fertility?

What is the fixed rule with regard to this?

What is the condition of the alkalies in most of their combinations? Of the acids?

What is said of phosphate of lime?

quire it in large quantities, it must be made soluble, which is done by combining it with an alkali.

For instance, if the silica in the soil is insoluble, we must make an application of an alkali, such as potash, which will unite with the silica, and form the silicate of potash, which is in the exact condition to be dissolved and carried into the roots of plants.

Chlorine in the soil is probably always in an available condition.

Oxide of iron exists, as has been previously stated, usually in the form of the *per*oxide (or red oxide). Sometimes, however, it exists in the form of the *prot*oxide (or black oxide), which is poisonous to plants, and renders the soil unfertile. By loosening the soil in such a manner as to admit air and water, this compound takes up more oxygen, which renders it a *per*oxide, and makes it available for plants. The oxide of manganese is probably of little consequence.

The usefulness of all of these matters in the soil depends on their *exposure ;* if they are in the *interior* of particles, they cannot be made use of; while, if the particles are so pulverized that their constituents are exposed, they become available, because water can immediately attack to dissolve, and carry them into roots.

How may silica be rendered soluble?
What is the condition of chlorine in the soil?
Do peroxide and protoxide of iron affect plants in the same way?
How would you treat a soil containing protoxide of iron?
On what does the usefulness of all these matters in the soil depend?

This is one of the great offices of plowing and hoeing; the *lumps* of soil being thereby more broken up and exposed to the action of atmospheric influences, which are often necessary to produce a fertile condition of soil, while the trituration of particles reduces them in size.

SUBSOIL.

The subsoil is usually of a different character from the surface soil, but this difference is more often the result of circumstances than of formation. The surface soil from having been long cultivated has been more opened to the influences of the air than is the case with the subsoil, which has never been disturbed so as to allow the same action. Again the growth of plants has supplied the surface soil with roots, which by decaying have given it organic matter, thus darkening its color, rendering it warmer, and giving greater ability to absorb heat and moisture, and to retain manures. All of these effects render the surface soil of a more fertile character than it was before vegetable growth commenced; and, where frequent cultivation and manures have been applied, a still greater benefit has resulted. In most instances the subsoil may by the same means

---

What is one of the chief offices of plowing and hoeing?
Is the subsoil usually different from the surface soil?
What circumstances have occasioned the difference?
In what way?

be gradually improved in condition until it equals the surface soil in fertility. The means of producing this result, also farther accounts of its advantages, will be given under the head of *Cultivation* (Sect. IV.)

IMPROVEMENT.

From what has now been said of the character of the soil, it must be evident that, as we know the *causes* of fertility and barrenness, we may by the proper means improve the character of all soils which are not now in the highest state of fertility.

Chemical analysis will tell us the *composition* of a soil, and an examination, such as any farmer may make, will inform us of its deficiencies in *mechanical* character, and we may at once resort to the proper means to secure fertility. In some instances the soil may contain every thing that is required, but not in the necessary condition. For instance, in some parts of Massachusetts, there are nearly *barren* soils which show by analysis precisely the same chemical composition as the soil of the Miami valley of Ohio, one of the most *fertile* in the world. The cause of this great difference in their agricultural capabilities, is that the Miami soil has its particles

---

May the subsoil be made to resemble the surface soil?
May all soils be brought to the highest state of fertility?
On what examination must improvement be based?
What is the difference between the soil of some parts of Massachusetts and that of the Miami valley?

finely pulverized; while in the Massachusetts soil the ingredients are combined within particles (such as pebbles, etc.), where they are out of the reach of roots.

In other cases, we find two soils, which are equally well pulverized, and which appear to be of the same character, having very different power to support crops. Chemical analysis will show in these instances a difference of composition.

All of these differences may be overcome by the use of the proper means. Sometimes it could be done at an expense which would be justified by the result; and, at others, it might require too large an outlay to be profitable. It becomes a question of economy, not of ability, and science is able to estimate the cost.

Soil cannot be cultivated understandingly until it has been subjected to such an examination as will tell us exactly what is necessary to render it fertile. Even after fertility is perfectly restored it requires thought and care to maintain it. The ingredients of the soil must be returned in the form of manures as largely as they are removed by the crop, or the supply will eventually become too small for the purposes of vegetation.

Why do soils of the same degree of fineness sometimes differ in fertility?

Can soils always be rendered fertile with profit?

Can we determine the cost before commencing the work?

What must be done before a soil can be cultivated understandingly?

What must be done to keep up the quality of the soil?

SECTION THIRD.

MANURES.

SECTION THIRD.

MANURES.

CHAPTER I.

CHARACTER AND VARIETIES OF MANURES.

To understand the science of *manures* is the most important branch of practical farming. No baker would be called a good practical baker who kept his flour exposed to the sun and rain. No shoemaker would be called a good practical shoemaker, who used morocco for the soles of his shoes, and heavy leather for the uppers. No carpenter would be called a good practical carpenter, who tried to build a house without nails, or other fastenings. So with the farmer. He cannot be called a good practical farmer if he keeps the materials, from which he is to make plants, in such a condition, that they will have their value destroyed, uses them in the wrong places, or tries to put them together without having every thing present that is necessary. Before he can avoid failures *with certainty*, he must know what manures are composed of, how they are to be preserved, where they are needed, and what kinds are required. True, he may from observation and experience, *guess* at results, but he cannot *know* that he is right until he has learned the facts above named. In this section of our work, we mean to convey some of the information necessary to this branch of *practical farming.*

We shall adopt a classification of the subject somewhat different from that found in most works on manures, but the *facts* are the same. The action of manures is either *mechanical* or *chemical*, or a combination of both. For instance: some kinds of manure improve the mechanical character of the soil, such as those which loosen stiff clay soils, or others which render light sandy soils compact—these are called *mechanical* manures. Some again furnish food for plants—these are called *chemical* manures.

Many mechanical manures produce their effects by means of chemical action. Thus *potash* combines chemically with sand in the soil. In so doing, it

---

What must a farmer know in order to avoid failures?
Can this be learned entirely from observation?
What kind of action have manures?
Give examples of each of these.
May mechanical effects be produced by chemical action?
How does potash affect the soil?

roughens the surfaces of the particles of sand, and renders the soil less liable to be compacted by rains. In this manner, it acts as a *mechanical* manure. The compound of sand and potash,* as well as the potash alone, may enter into the composition of plants, and hence it is a *chemical* manure. In other words, potash belongs to both classes described above.

It is important that this distinction should be well understood by the learner, as the words "mechanical" and "chemical" in connection with manures will be made use of throughout the following pages.

There is another class of manures which we shall call *absorbents*. These comprise those substances which have the power of taking up fertilizing matters, and retaining them for the use of plants. For instance, *charcoal* is an absorbent. As was stated in the section on soils, this substance is a retainer of all fertilizing gases and many minerals. Other matters made use of in agriculture have the same effect. These absorbents will be spoken of more fully in their proper places.

TABLE.

MECHANICAL MANURES	are those which improve the mechanical condition of soils.
CHEMICAL "	are those which serve as food for plants.

* Silicate of potash.

What are absorbents?
What kind of manure is charcoal?

ABSORBENTS are those substances which absorb and retain fertilizing matters.

Manures may be divided into three classes, viz.: *organic*, *inorganic*, and *atmospheric*.

ORGANIC manures comprise all *animal* and *vegetable* matters which are used to fertilize the soil, such as dung, muck, etc.

INORGANIC manures are those which are of a purely *mineral* character, such as lime, ashes, etc.

ATMOSPHERIC manures consist of those organic manures which are in the form of gases in the atmosphere, and which are absorbed by rains and carried to the soil. These are of immense importance. The ammonia and carbonic acid in the air are atmospheric manures.

CHAPTER II.

EXCREMENTS OF ANIMALS.

THE first organic manure which we shall examine, is animal *excrement*.

This is composed of those matters which have been eaten by the animal as food, and have been thrown off as solid or liquid manure. In order that

---

Into what classes may manures be divided?
What are organic manures?
Inorganic? Atmospheric?

we may know of what they consist, we must refer to the composition of food and examine the process of digestion.

The food of animals, we have seen to consist of both organic and inorganic matter. The organic part may be divided into two classes, *i. e.*, that portion which contains nitrogen—such as gluten, albumen, etc., and that which does not contain nitrogen —such as starch, sugar, oil, etc.

The inorganic part of food may also be divided into *soluble* matter and *insoluble* matter.

DIGESTION AND ITS PRODUCTS.

Let us now suppose that we have a full-grown ox, which is not increasing in any of his parts, but only consumes food to keep up his respiration, and to supply the natural wastes of his body. To this ox we will feed a ton of hay which contains organic matter, with and without nitrogen, and soluble and insoluble inorganic substances. Now let us try to follow it through its changes in the animal, and observe its destination. Liebig compares the consumption of food by animals to the imperfect burning of wood in a stove, where a portion of the fuel is resolved into gases and ashes (that is, it is completely burned), and

---

Of what is animal excrement composed?
Explain the composition of the food of animals.
What does hay contain?
To what does Liebig compare the consumption of food by animals, and why?

another portion, which is not thoroughly burned, passes off as *soot*. In the animal action in question, the food undergoes changes which are similar to this burning of wood. A part of the food is *digested* and taken up by the blood, while another portion remains undigested, and passes the bowels as solid dung—corresponding to soot. This part of the dung then, we see is merely so much of the food as passes through the system without being materially changed. Its nature is easily understood. It contains organic and inorganic matter in nearly the same condition as they existed in the hay. They have been rendered finer and softer, but their chemical character is not materially altered. The dung also contains small quantities of nitrogenous matter, which *leaked out*, as it were, from the stomach and intestines. The digested food, however, undergoes further changes which affect its character, and it escapes from the body in three ways—*i. e.*, through the lungs, through the bladder, and through the bowels. It will be recollected from the first section of this book, p. 22, that the carbon in the blood of animals, unites with the oxygen of the air drawn into the lungs, and is thrown off in the breath as carbonic acid. The hydrogen and oxygen unite to form a part of the water which constitutes the moisture of the breath.

Of what does that part of dung consist which resembles soot?
What else does the dung contain?
In what manner does the digested part of food escape from the body?

That portion of the organic part of the hay which has been taken up by the blood of the ox, and which does not contain nitrogen (corresponding to the *first* class of proximates, as described in Sect. I), is emitted through the lungs. It consists, as will be recollected, of carbon, hydrogen and oxygen, and these assume, in respiration, the form of carbonic acid and water.

The organic matter of the digested hay, in the blood, which contains nitrogen (corresponding to the *second* class of proximates, described in Sect. I), goes to the *bladder*, where it assumes the form of urea—a constituent of urine or liquid manure.

We have now disposed of the imperfectly digested food (dung), and of the *organic* matter which was taken up by the blood. All that remains to be examined is the inorganic or mineral matter in the blood, which would have become *ashes*, if the hay had been burned. The *soluble* part of this inorganic matter passes into the bladder, and forms the *inorganic part of urine*. The *insoluble* part passes the bowels, in connection with the dung.

If any of the food taken up by the blood is not returned as above stated, it goes to form fat, muscle, hair, bones, or some other part of the animal, and as

---

Explain the escape of carbon, hydrogen and oxygen.
What becomes of the nitrogenous parts?
How is the *soluble* ash of the digested food parted with?
The insoluble?
If any portions of the food are not returned in the dung, how are they disposed of?

he is not growing (not increasing in weight) an equivalent amount of the body of the animal goes to the manure to take the place of the part retained.*

We now have our subject in a form to be readily understood. We learn that when food is given to animals it is not *put out of existence,* but is merely *changed in form ;* and that in the impurities of the breath, we have a large portion of those parts of the food which plants obtain from air and from water; while the solid and liquid excrements contain all that was taken by the plants from the soil and manures.

The SOLID DUNG contains	the undigested parts of the food, the *insoluble* parts of the ash, and the nitrogenous matters which have *escaped* from the digestive organs.
" LIQUID MANURE "	the nitrogenous or *second class* of proximates of the digested food, and the *soluble* parts of the ash.

* This account of digestion is not, perhaps, strictly accurate in a physiological point of view, but it is sufficiently so to give an elementary understanding of the character of excrements as manures.

How is their place supplied?
Is food put out of existence when it is fed to animals?
What does the solid dung contain? Liquid manure? The breath?

The BREATH contains the *first class* of proximates, those which contain carbon, hydrogen and oxygen, but *no nitrogen.**

CHAPTER III.

WASTE OF MANURE.

THE loss of manure is a subject which demands most serious attention. Until within a few years, little was known about the true character of manures, and consequently, of the importance of protecting them against loss.

The first causes of waste are *evaporation* and *leaching*.

EVAPORATION.

Evaporation is the changing of a solid or liquid body to a vapory form. Thus common smelling salts, a solid, if left exposed, passes into the atmosphere in

* The excrements of animals contain more or less of sulphur, and sometimes small quantities of phosphorus.

What are the first causes of loss of manure?
What is *evaporation?*

the form of a gas or vapor. Water, a liquid, evaporates, and becomes a vapor in the atmosphere. This is the case with very many substances, and in organic nature, both solid and liquid, they are liable to assume a gaseous form, and become mixed with the atmosphere. They are not destroyed, but are merely changed in form.

As an instance of this action, suppose an animal to die and to decay on the surface of the earth. After a time, the flesh will entirely disappear, but is not lost. It no longer exists as the flesh of an animal, but its carbon, hydrogen, oxygen, and nitrogen, still exist in the air. They have been liberated from the attractions which held them together, and have passed away; but (as we already know from what has been said in a former section) they are ready to be again taken up by plants, and pressed into the service of life.

The evaporation of liquids may take place without the aid of any thing but heat; still, in the case of solids, it is often assisted by decay and combustion, which break up the bonds that hold the constituents of bodies together, and thus enable them to return to the atmosphere, from which they were originally derived.

It must be recollected that every thing, which has

Name a solid body which evaporates.

What takes place when a dead animal is exposed to the atmosphere for a sufficient time?

What often assist the evaporation of solids?

an *odor* (or can be smelled), is evaporating. The odor is caused by parts of the body floating in the air, and acting on the nerves of the nose. This is an invariable rule ; and, when we perceive an odor, we may be sure that parts of the material, from which it emanates, are escaping. If we perceive the odor of an apple, it is because parts of the volatile oils of the apple enter the nose. The same is true when we smell hartshorn, cologne, etc.

Manures made by animals have an offensive odor, simply because volatile parts of the manure escape into the air, and are therefore made perceptible. All organic parts in turn become volatile, assuming a gaseous form as they decompose.

We do not see the gases rising, but there are many ways by which we can detect them. If we wave a feather over a manure heap, from which ammonia is escaping, the feather having been recently dipped in manure, white fumes will appear around the feather, being the muriate of ammonia formed by the union of the escaping gas with the muriatic acid. Not only ammonia, but also carbonic acid, and other gases which are useful to vegetation escape, and are given to the winds. Indeed it may be stated in few words that all of the organic part of *plants* (all that was obtained from the air, water, and ammonia),

---

What is the cause of odor?
When we perceive an odor, what is taking place?
Why do manures give off offensive odors?
How may we detect ammonia escaping from manure?

constituting more than nine tenths of their dry weight, may be evaporated by the assistance of decay or combustion. The organic part of *manures* may be lost in the same manner; and, if the process of decomposition be continued long enough, nothing but a mass of mineral matter will remain, except perhaps a small quantity of carbon which has not been resolved into carbonic acid.

The proportion of solid manure lost by evaporation (made by the assistance of decay), is a very large part of the whole. Manure cannot be kept a single day in its natural state without losing something. It commences to give out an offensive odor immediately, and this odor is occasioned, as was before stated, by the loss of some of its fertilizing parts.

Animal manure contains, as will be seen by reference to p. 100, all of the substances contained in plants, though not always in the correct relative proportions to each other. When decomposition commences, the carbon unites with the oxygen of the air, and passes off as carbonic acid; the hydrogen and oxygen combine to form water (which evaporates), and the *nitrogen is mostly resolved into ammonia, which escapes into the atmosphere.*

What remains after manure has been long exposed to decomposition?

What gaseous compounds are formed by the decomposition of manures?

If manure is thrown into heaps, it often ferments so rapidly as to produce sufficient heat to set fire to some parts of the manure, and cause it to be thrown off with greater rapidity. This may be observed in nearly all heaps of animal excrement. When they have lain for some time in mild weather, gray streaks of *ashes* are often to be seen in the centre of the pile. The organic part of the manure having been *burned* away, nothing but the ash remains,—this is called *fire-fanging.*

Manures kept in cellars without being mixed with refuse matter are subject to the same losses.

When kept in the yard, they are still liable to be lost by evaporation. They are here often saturated with water, and this water in its evaporation carries away the ammonia, and carbonic acid which it has obtained from the rotting mass. The evaporation of the water is rapidly carried on, on account of the great extent of surface. The whole mass is spongy, and soaks the liquids up from below (through hollow straws, etc.), to be evaporated at the surface on the same principle as causes the wick of a lamp to draw up the oil to supply fuel for the flame.

Liquid Manure containing large quantities of nitrogen, and forming much ammonia, is also liable to lose all of its organic part from evaporation (and

---

Describe fire-fanging.

What takes place when animal manure is exposed in an open barn-yard?

What does liquid manure lose by evaporation?

fermentation), so that it is rendered as much less valuable as is the solid dung.*

From these remarks, it may be justly inferred that a very large portion of the *value* of solid and liquid manure as ordinarily kept is lost by evaporation in a sufficient length of time, depending on circumstances, whether it be three months or several years. The wasting commences as soon as the manure is dropped, and continues, except in very cold weather, until the destruction is complete. Hence we see that true economy requires that the manures of the stable, stye, and poultry-house, should be protected from evaporation (as will be hereafter described), as soon as possible after they are made.

LEACHING.

The subject of *leaching* is as important in considering the *inorganic* parts of manures as evaporation is to the organic, while leaching also affects the organic gases, they being absorbed by water in a great degree.

A good illustration of leaching is found in the manufacture of potash. When water is poured

* It should be recollected that every bent straw may act as a syphon, and occasion much loss of liquid manure.

When does the waste of exposed manure commence?
What does economy of manure require?
What is the effect of leaching?
Give an illustration of leaching.

over wood-ashes, it dissolves their potash which it carries through in solution, making ley. If ley is boiled to dryness, it leaves the potash in a solid form, proving that this substance had been dissolved by the water and removed from the insoluble parts of the ashes.

In the same way water in passing through manures takes up the soluble portions of the ash as fast as liberated by decomposition, and carries them into the soil below; or, if the water runs off from the surface, they accompany it. In either case they are lost to the manure. There is but a small quantity of ash exposed for leaching in recent manures; but, as the decomposition of the organic part proceeds, it continues to develope it more and more (in the same manner as burning would do, only slower), thus preparing fresh supplies to be carried off with each shower. In this way, while manures are largely injured by evaporation, the soluble inorganic parts are removed by water until but a small remnant of its original fertilizing properties remains.

It is a singular fact concerning leaching, that water is able to carry no part of the organic constituents of vegetables more than about thirty-four inches below the surface in a fertile soil. They would probably be carried to an unlimited distance

---

How does water affect decomposing manures?

Does continued decomposition continue to prepare material to be leached away?

How far from the surface of the soil may organic constituents be carried by water?

in pure sand, as it contains nothing which is capable of arresting them ; but, in most soils, the clay and carbon which they contain retain all of the ammonia ; also nearly all of the matters which go to form the inorganic constituents of plants within about the above named distance from the surface of the soil. If such were not the case, the fertility of the earth must soon be destroyed, as all of those elements which the soil must supply to growing plants would be carried down out of the reach of roots, and leave the world a barren waste, its surface having lost its elements of fertility, while the downward filtration of these would render the water of wells unfit for our use. Now, however, they are all retained near the surface of the soil, and the water issues from springs comparatively pure.

EVAPORATION removes from manure—

Carbon, in the form of carbonic acid.
Hydrogen and oxygen, in the form of water.
Nitrogen, in the form of ammonia.

LEACHING removes from manure—

The soluble and most valuable parts of the ash in solution in water, besides carrying away some of the named above forms of organic matter.

What arrests their farther progress?
What would be the effect of allowing these matters to filter downwards?
What does evaporation remove from manure? Leaching?

CHAPTER IV.

ABSORBENTS.

BEFORE considering farther the subject of animal excrement, it is necessary to examine a class of manures known as *absorbents.* These comprise all matters which have the power of absorbing, or soaking up, as it were, the gases which arise from the evaporation of solid and liquid manures, and retaining them until required by plants.

The most important of these is undoubtedly *carbon* or charcoal.

CHARCOAL.

Charcoal, in an agricultural sense, means all forms of carbon, whether as peat, muck, charcoal dust from the spark-catchers of locomotives, charcoal hearths, river and swamp deposits, leaf mould, decomposed spent tanbark or sawdust, etc. In short, if any vegetable matter is decomposed with the partial exclusion of air (so that there shall not be oxygen enough supplied to unite with all of the carbon), a

---

What substances are called absorbents?
What is the most important of these?
What substances are called charcoal in agriculture?
How is vegetable matter rendered useful as charcoal?

portion of its carbon remains in the exact condition to serve the purposes of charcoal.

The offices performed in the soil by carbonaceous matter were fully explained in a former section (p. 79, Sect. 2), and we will now examine merely its action with regard to manures. When properly applied to manures, in compost, it has the following effects :

1. It absorbs and retains the fertilizing gases evaporating from decomposing matters.

2. It acts as a *divisor*, thereby reducing the strength (or intensity) of powerful manures—thus rendering them less likely to injure the roots of plants ; and also increases their bulk, so as to prevent *fire fanging* in composts.

3. It in part prevents the leaching out of the soluble parts of the ash.

4. It keeps the compost moist.

The first-named office of charcoal, *i. e.*, absorbing and retaining gases, is one of the utmost importance. It is this quality that gives to it so high a position in the opinion of all who have used it. As was stated in the section on soils, carbonaceous matter seems to be capable of absorbing every thing which may be of use to vegetation. It is a grand purifier, and while it prevents offensive odors from escaping, it is at the same time storing its pores with food for the nourishment of plants.

What is the first-named effect of charcoal? The second? Third? Fourth?

Explain the first action.

2d. In its capacity as a *divisor* for manures, charcoal should be considered as excellent in all cases, especially to use with strongly concentrated (or heating) animal manures. These, when applied in their natural state to the soil, are very apt to injure young roots by the violence of their action. When mixed with a divisor, such manures are *diluted*, made less active, and consequently less injurious. In composts, manures are liable, as has been before stated, to become burned by the resultant heat of decomposition; this is called *fire fanging*, and is prevented by the liberal use of divisors, because, by increasing the bulk, the heat being diffused through a larger mass, becomes less intense. The same principle is exhibited in the fact that it takes more fire to boil a cauldron of water than a tea-kettle full.

3d. Charcoal has much power to arrest the passage of mineral matters in solution; so much so, that compost heaps, well supplied with muck, are less affected by rains than those not so supplied. All composts, however, should be kept under cover.

4th. Charcoal keeps the compost moist from the ease with which it absorbs water, and its ability to withstand drought.

With these advantages before us, we must see the importance of an understanding of the modes for

---

Explain its action as a divisor.

How does charcoal protect composts against injurious action of rains?

How does it keep them moist?

obtaining charcoal. Many farmers are so situated that they can obtain sufficient quantities of charcoal dust. Others have not equal facilities. Nearly all, however, can obtain *muck*, and to this we will now turn our attention.

MUCK, AND THE LIME AND SALT MIXTURE.

By *muck*, we mean the vegetable deposits of swamps and rivers. It consists of decayed organic substances, mixed with more or less earth. Its principal constituent is *carbon*, in different degrees of development, which has remained after the decomposition of vegetable matter. Muck varies largely in its quality, according to the amount of carbon which it contains, and the perfection of its decomposition. The best muck is usually found in comparatively dry locations, where the water which once caused the deposit has been removed. Muck which has been long in this condition, is usually better decomposed than that which is saturated with water. The muck from swamps, however, may soon be brought to the best condition. It should be thrown out, if possible, at least one year before it is required for use (a less time may suffice, except in very cold

---

What source of carbon is within the reach of most farmers?
What do we mean by muck?
Of what does it consist?
How does it differ in quality?

climates) and left, in small heaps or ridges, to the action of the weather, which will assist in pulverizing it, while, from having its water removed, its decomposition goes on more rapidly.

After the muck has remained in this condition a sufficient length of time, it may be removed to the barn-yard and composted with the lime and salt mixture (described on page 115) in the proportion of one cord of muck to four bushels of the mixture. This compost ought to be made under cover, lest the rain leach out the constituents of the mixture, and thus occasion loss ; at the end of a month or more, the muck in the compost will have been reduced to a fine pulverulent mass, nearly equal to charcoal dust for application to animal excrement. When in this condition it is called *prepared* muck, by which name it will be designated in the following pages.

Muck should not be used immediately after being taken from the swamp, as it is then almost always *sour*, and is liable to produce sorrel. Its *sourness* is due to *acids* which it contains, and these must be rectified by the application of an alkali, or by long exposure to the weather, before the muck is suitable for use.

What is the first step in preparing muck for decomposition?
With what proportion of the lime and salt mixture should it be composted?
Why should this compost be made under cover?
What is this called after decomposition?
Why should we not use muck immediately after taking it from the swamp?

LIME AND SALT MIXTURE.

The lime and salt mixture, used in the decomposition of muck, is made in the following manner:

RECIPE.—Take *three* bushels of shell lime, *hot from the kiln,* or as fresh as possible, and slake it with water in which *one* bushel of salt has been dissolved.

Care must be taken to use only so much water as is necessary to dissolve the salt, as it is difficult to induce the lime to absorb a larger quantity.

In dissolving the salt, it is well to hang it in a basket in the upper part of the water, as the salt water will immediately settle towards the bottom (being heavier), and allow the freshest water to be nearest to the salt. In this way, the salt may be all dissolved, and thus make the brine used to slake the lime. It may be necessary to apply the brine at intervals of a day or two, and to stir the mass often, as the amount of water is too great to be readily absorbed.

This mixture should be made under cover, as, if exposed, it would obtain moisture from rain or dew, which would prevent the use of all the brine.

What proportions of lime and salt are required for the decomposing mixture?

Explain the process of making it.

Why should it be made under cover?

Another objection to its exposure to the weather is its great liability to be washed away by rains. It should be at least ten days old before being used, and would probably be improved by an age of three or four months, as the chemical changes it undergoes will require some time to be completed.

The character of this mixture may be best described by the following diagram :—

We have originally—

Lime — Salt consisting of

Chlorine and Sodium. } Chloride of Sodium.

—Carbonic acid and —Oxygen in the air.

Chloride of lime. . Carbonate of Soda.

*

The lime unites with the chlorine of the salt and forms *chloride of lime.*

The sodium, after being freed from the chlorine, unites with the oxygen of the air and forms soda,

* There is, undoubtedly, some of this lime which does not unite with the chlorine; this, however, is still as valuable as any lime.

Explain the character of this mixture as represented in the diagram. (Black board.)

which, combining with the carbonic acid of the atmosphere, forms carbonate of soda.

Chloride of lime and carbonate of soda are better agents in the decomposition of muck than pure salt and lime ; and, as these compounds are the result of the mixture, much benefit ensues from the operation.

When *shell* lime cannot be obtained, Thomaston, or any other very pure lime, will answer, though care must be taken that it do not contain much magnesia.

LIME.

Muck may be decomposed by the aid of other materials. *Lime* is very efficient, though not as much so as when combined with salt. The action of lime, when applied to the muck, depends very much on its condition. Air-slaked lime (carbonate of lime), and hydrate of lime, slaked with water, have but a limited effect compared with lime freshly burned and applied in a caustic (or pure) form. When so used, however, the compost should not be exposed to rains, as this would have a tendency to make *mortar* which would harden it.

What effect has lime on muck?

On what does the energy of this effect depend?

Why should a compost of muck and lime be protected from rain?

POTASH.

Potash is a very active agent in decomposing vegetable matter, and may be used with great advantage, especially where an analysis of the soil which is to be manured shows a deficiency of potash.

Unleached wood ashes are generally the best source from which to obtain this, and from five to twenty-five bushels of these mixed with one cord of muck will produce the desired result.*

The sparlings (or refuse) of potash warehouses may often be purchased at sufficiently low rates to be used for this purpose, and answer an excellent end. They may be applied at the rate of from twenty to one hundred pounds to each cord of muck.

By any of the foregoing methods, muck may be *prepared* for use in composting.

* *Leached* ashes will not supply the place of these, as the leaching has deprived them of their potash.

Is potash valuable for this use?
From what sources may potash be obtained?
In what proportion should ashes be applied to muck? Sparlings?

CHAPTER V.

COMPOSTING STABLE MANURE.

In composting stable manure in the most economical manner, the evaporation of the organic parts and the leaching of the ashy (and other) portions must be avoided, while the condition of the mass is such as to admit of the perfect decomposition of the manure.

Solid manures in their fresh state are of but very little use to plants. It is only as they are decomposed, and have their nitrogen turned into ammonia, and their other ingredients resolved into the condition required by plants, that they are of much value as fertilizers. We have seen that, if this decomposition takes place without proper precautions being made, the most valuable parts of the manure would be lost. Nor would it be prudent to keep manures from decomposing until they are applied to the soil, for then they are not immediately ready for use, and time is lost. By composting, we aim to save every thing while we prepare the manures for immediate use.

What principles should regulate us in composting?
In what condition is solid dung of value as a fertilizer?
What do we aim to do in composting?

SHELTER.

The first consideration in preparing for composting, is to provide proper shelter. This may be done either by means of a shed or by arranging a cellar under the stables, or in any other manner that may be dictated by circumstances. It is no doubt better to have the manure shed enclosed so as to make it an effectual protection ; this however is not absolutely necessary if the roof project far enough over the compost to shelter it from the sun's rays and from driving rains.

The importance of some protection of this kind, is evident from what has already been said, and indeed it is impossible to make an economical use of manures without it. The trifling cost of building a shed, or preparing a cellar, is amply repaid in the benefit resulting from their uses.

THE FLOOR.

The *floor* or foundation on which to build the compost deserves some consideration. It may be of plank tightly fitted, a hard bed of clay, or better, a cemented surface. Whatever material is used in its construction (and stiff clay mixed with water and

---

What is the first consideration for composts?
Describe the arrangement of floor.

beaten compactly down answers an excellent purpose), the floor must have such an inclination as will cause it to discharge water only at one point. That is, one part of the edge must be lower than the rest of the floor, which must be so shaped that water will run towards this point from every part of it; then—the floor being water-tight—all of the liquids of the compost may be collected in a

TANK.

This *tank* used to collect the liquids of the manure may be made by sinking a barrel or hogshead (according to the size of the heap) in the ground at the point where it is required, or in any other convenient manner.

In the tank a pump of cheap construction may be placed, to raise the liquid to a sufficient height to be conveyed by a trough to the centre of the heap, and there distributed by means of a perforated board with raised edges, and long enough to reach across the heap in any direction. By altering the position of this board, the liquid may be carried evenly over the whole mass.

The appearance of the apparatus required for composting, and the compost laid up, may be better shown by the following figure.

How should the tank be attached?

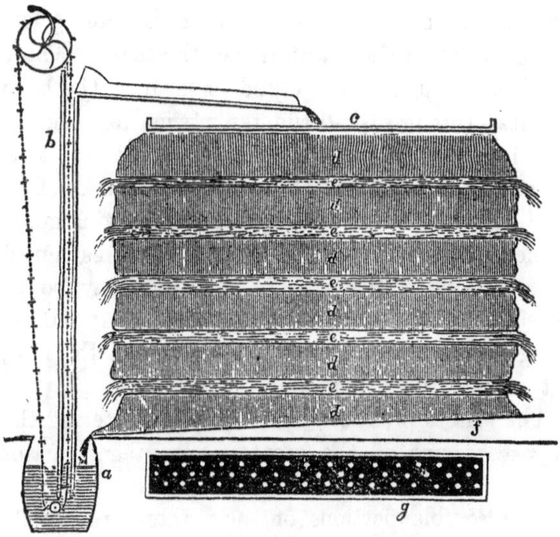

Fig. 2.

a, tank; *b*, pump; *c* & *g*, perforated board; *d*, muck; *e*, manure; *f*, floor.

The compost is made by laying on the floor ten or twelve inches of muck, and on that a few inches of manure, then another heavy layer of muck, and another of manure, continuing in this manner until the heap is raised to the required height, always having a thick layer of muck at the top.

After laying up the heap, the tank should be filled with liquid manure from the stables, slops from

How is the compost made?

the house, soap-suds, or other water containing fertilizing matter, to be pumped over the mass. There should be enough of the liquid to saturate the heap and filter through to fill the tank twice a week, at which intervals it should be again pumped up, thus continually being passed through the manure. This liquid should not be changed, as it contains much soluble manure. Should the liquid manures named above not be sufficient, the quantity may be increased by the use of rain-water. That falling during the first ten minutes of a shower is the best, as it contains much ammonia.

The effects produced by frequently watering the compost is one of the greatest advantages of this system.

The soluble portions of the manure are equally diffused through every part of the heap.

Should the heat of fermentation be too great, the watering will reduce it.

When the compost is saturated with water, the air is driven out ; and, as the water subsides, *fresh* air enters and takes its place. This fresh air contains oxygen, which assists in the decomposition of the manure.

In short, the watering does all the work of forking over by hand much better and much more cheaply.

What liquids are best for moistening the compost?
How should they be applied?
What are the advantages of this moistening?
How does it compare with forking over?

At the end of a month or more, this compost will be ready for use. The layers in the manure will have disappeared, the whole mass having become of a uniform character, highly fertilizing, and ready to be immediately used by plants.

It may be applied to the soil, either as a top-dressing, or otherwise, without fear of loss, as the muck will retain all of the gases which would otherwise evaporate.

The cost and trouble of the foregoing system of composting are trifling compared with its advantages. The quantity of the manure is much increased, and its quality improved. The health of the animals is secured by the retention of those gases, which, when allowed to escape, render impure the air which they have to breathe.

The cleanliness of the stable and yard is much advanced as the effete matters, which would otherwise litter them, are carefully removed to the compost.

As an instance of the profit of composting, it may be stated that Prof. Mapes has decomposed ninety-two cords of swamp muck, with four hundred bushels of the lime and salt mixture, and then composted it with eight cords of *fresh* horse dung, making one hundred cords of manure fully equal to the same amount of stable-manure alone, which has lain one

---

Why will the ammonia of manure thus made, not escape if it be used as a top dressing?

What are the advantages of preparing manures in this manner?

What is the profit attending it?

year exposed to the weather. Indeed one cord of muck well decomposed, and containing the chlorine lime and soda of four bushels of the mixture, is of itself equal in value to the same amount of manure which has lain in an open barn-yard during the heat and rain of one season, and is then applied to the land in a *raw* or undecomposed state.

The foregoing system of composting is the best that has yet been suggested for making use of solid manures. Many other methods may be adopted when circumstances will not admit of so much attention. It is a common and excellent practice to throw prepared muck into the cellar under the stables, to be mixed and turned over with the manure by swine. In other cases the manures are kept in the yard, and are covered with a thin layer of muck every morning. The principle which renders these systems beneficial is the absorbent power of charcoal.

LIQUID MANURE.

Liquid manure from animals may, also, be made useful by the assistance of prepared muck. Where a tank is used in composting, the liquids from the stable may all be employed to supply moisture to the heap; but where any system is adopted, not requir-

---

In what other manners may muck be used in the preservation of manures?

How may liquid manure be made most useful?

ing liquids, the urine may be applied to muck heaps, and then allowed to ferment. Fermentation is necessary in urine as well as in solid dung, before it is very active as a manure. Urine, as will be recollected, contains nitrogen and forms ammonia on fermentation.

It is a very good plan to dig out the bottoms of the stalls in a circular or gutter-like form, three or four feet deep in the middle, cement the ground, or make it nearly water-tight, by a plastering of stiff clay, and fill them up with prepared muck. The appearance of a cross section of the floor thus arranged would be as follows :

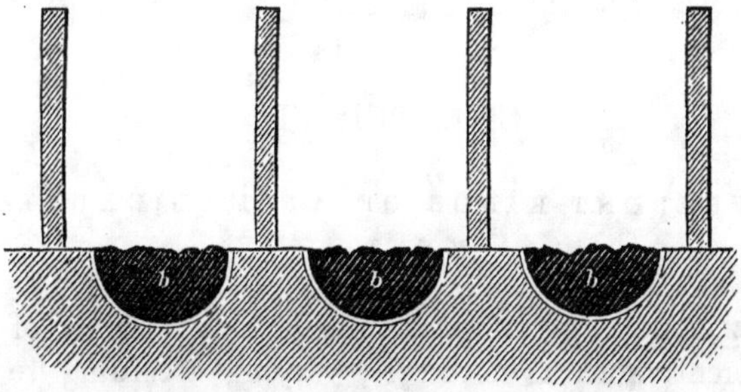

Fig. 3.

The prepared muck in the bottom of the stalls would absorb the urine as soon as voided, while yet warm with the animal heat, and receive heat from the animal's body while lying down at night. This

Describe the manner of digging out the bottoms of stalls.

heat will hasten the decomposition of the urea,* and if the muck be renewed twice a month, and that which is removed composted under cover, it will be found a most prolific source of good manure. In Flanders, the liquid manure of a cow is considered worth $10 per year, and it is not less valuable here. As was stated in the early part of this section, the inorganic (or mineral) matter contained in urine, is soluble, and consequently is immediately useful as food for plants.

By referring to the analysis of liquid and solid manure. in section V., their relative value may be seen.

CHAPTER VI.

DIFFERENT KINDS OF ANIMAL EXCREMENT.

The manures of different animals are, of course, of different value, as fertilizers, varying according to the food, the age of the animals, etc.

STABLE MANURE.

By stable manure we mean, usually, that of the

* The nitrogenous compound in the urine.

horse, and that of horned cattle. The case described in chap. 2 (of this section), was one where the animal was not increasing in any of its parts, but returned, in the form of manure, and otherwise, the equivalent of every thing eaten.. This case is one of the most simple kind, and is subject to many modifications.

The *growing* animal is increasing in size, and as he derives his increase from his food, he does not return in the form of manure as much as he eats. If his bones are growing, he is taking from his food phosphate of lime and nitrogenous matter; consequently, the manure will be poorer in these ingredients. The same may be said of the formation of the muscles, in relation to nitrogen.

The *fatting* animal, if full grown, makes manure which is as good as that from animals that are not increasing in size, because the fat is taken from those parts of the food which is obtained by plants from the atmosphere, and from nature, (*i. e.* from the 1st class of proximates). Fat contains no nitrogen, and, consequently, does not lessen the amount of this ingredient in the manure.

Milch Cows turn a part of their food to the for-

Is the manure of full-grown animals of the same quality as that of other animals?

Why does that of the growing animal differ?

Why does not the formation of *fat* reduce the quality of manure?

What does *milk* remove from the food?

mation of milk, and consequently, they produce manure of reduced value.

The solid manure of the horse is better than that of the ox, while the liquid manure of the ox is comparatively better than that of the horse. The cause of this is that the horse has poorer digestive organs than the ox, and consequently passes more of the valuable parts of his food, in an undigested form, as dung, while the ox, from chewing the cud and having more perfect organs, turns more of his food into urine than the horse.

RECAPITULATION.

FULL GROWN animals not producing milk, and full grown animals fattening } make the best manure.

GROWING ANIMALS reduce the value of their manure, taking portions of their food to form their bodies.

MILCH COWS reduce the value of their manure by changing a part of their food into milk.

THE OX makes poor dung and rich urine.

THE HORSE makes rich dung and poor urine.*

* Comparatively.

How do the solid and liquid manure of the horse and ox compare?

What occasions these differences?

NIGHT SOIL.

The *best* manure within the reach of the farmer is *night soil,* or human excrement. There has always been a false delicacy about mentioning this fertilizer, which has caused much waste, and great loss of health, from the impure and offensive odors which it is allowed to send forth to taint the air.

The value of the night soil yearly lost in the United States is, probably, about *fifty millions of dollars* (50,000,000) ; an amount nearly equal to the entire expenses of our National Government. Much of the ill health of our people is undoubtedly occasioned by neglecting the proper treatment of night soil.

That which directly affects agriculture, as treated of in this book, is the value of this substance as a fertilizer. The manure of man consists (as is the case with that of other animals) of those parts of his food which are not retained in the increase of his body. If he be *growing,* his manure is poorer, as in the case of the ox, and it is subject to all the other modifications named in the early part of this chapter. His food is usually of a varied character, and is rich in nitrogen, the phosphates, and

---

What is the most valuable manure accessible to the farmer?

What is the probable value of the night soil yearly lost in the United States?

Of what does the manure of man consist?

other inorganic constituents ; consequently, his manure is made valuable by containing large quantities of these matters. As is the case with the ox, the *dung* contains the undigested food, the secretions (or leakings) of the digestive organs, and the insoluble parts of the ash of the digested food. The *urine*, in like manner, contains a large proportion of the nitrogen and the soluble inorganic parts of the digested food. When we consider how much richer the *food* of man is than that of horned cattle, we shall see the superior value of his *excrement.*

Night soil has been used as a manure, for ages, in China, which is, undoubtedly, one great secret of their success in supporting a dense population, for so long a time, without impoverishing the soil. It has been found, in many instances, to increase the productive power of the natural soil three-fold. That is, if a soil would produce ten bushels of wheat per acre, without manure, it would produce thirty bushels if manured with night soil.

Some have supposed that manuring with night soil would give disagreeable properties to plants: such is not the case ; their quality is invariably improved. The color and odor of the rose become richer and more delicate by the use of the most offensive night soil as manure.

Describe this manure as compared with the excrements of other animals.

Does the use of night soil produce disagreeable properties in plants?

It is evident that this is the case from the fact that plants have it for their direct object to make over and put together the refuse organic matter, and the gases and the minerals found in nature, for the use of animals. If there were no natural means of rendering the excrement of animals available to plants, the earth must soon be shorn of its fertility, as the elements of growth when once consumed would be essentially destroyed, and no soil could survive the exhaustion. There is no reason why the manure of man should be rejected by vegetation more than that of any other animal; and indeed it is not, for ample experience has proved that for most soils there is no better manure in existence.

A single experiment will suffice to show that night soil may be so kept that there shall be no loss of its valuable gases, and consequently no offensive odor arising from it, while it may be removed and applied to crops without unpleasantness. All that is necessary to effect this wonderful change in night soil, and to turn it from its disagreeable character to one entirely inoffensive, is to mix with it a little charcoal dust, prepared muck, or any other good absorbent—thus making what is called poudrette. The mode of doing this must depend on circumstances. In many cases, it would be expedient to

---

What is the direct object of plants?
What would result if this were not the case?
How may night soil be easily prepared for use, and its offensive odor prevented?

keep a barrel of the absorbent in the privy and throw down a small quantity every day. The effect on the odor of the house would amply repay the trouble.

The manure thus made is of the most valuable character, and may be used under any circumstances with a certainty of obtaining a good crop. It should not be used unmixed with some absorbent, as it is of such strength as to kill plants.

For an analysis of human manure, see Section V.

HOG MANURE.

Hog Manure is very valuable, but it must be used with care. It is so violent in its action that, when applied in a pure form to crops, it often produces injurious results. It is liable to make cabbages *clump-footed*, and to induce a disease in turnips called *ambury* (or fingers and toes). The only precaution necessary is to supply the stye with prepared muck, charcoal-dust, leaf-mould, or any absorbent in plentiful quantities, often adding fresh supplies. The hogs will work this over with the manure ; and, when required for use, it will be found an excellent fertilizer. The absorbent will have overcome its injurious tendency, and it may be safely applied to any crop. From the variety and rich character of the food of this animal, his manure is of a superior quality.

Should pure night soil be used as a manure?
What precaution is necessary in preparing hog manure for use?

Butchers' hog-pen manure is one of the best fertilizers known. It is made by animals that live almost entirely on blood and other animal refuse, and is very rich in nitrogen and the phosphates. It should be mixed with prepared muck, or its substitute, to prevent the loss of its ammonia, and as a protection against its injurious effect on plants.

POULTRY HOUSE MANURE.

Next in value to night soil, among domestic manures, are the excrements of poultry, pigeons, etc. Birds live on the nice bits of creation, seeds, insects, etc., and they discharge their solid and liquid excrements together. Poultry-dung is nearly equal in value to guano (except that it contains more water), and it deserves to be carefully preserved and judiciously used. It is as well worth twenty-five cents per bushel as guano is worth fifty dollars a ton (at which price it is now sold).

Poultry-manure is liable to as much injury from evaporation and leaching as is any other manure, and equal care should be taken (by the same means) to prevent such loss. Good shelter over the roosts, and daily sprinkling with prepared muck or charcoal-dust will be amply repaid by the increased value of

---

Why is the manure from butchers' hog-pens very valuable?
How does the value of poultry manure compare with that of guano?
How may it be protected against loss?

the manure, and its better action and greater durability in the soil. The value of this manure should be taken into consideration in calculating the profit of keeping poultry (as indeed with all other stock). It has been observed by a gentleman of much experience, in poultry raising, that the yearly manure of a hundred fowls applied to previously unmanured land would produce *extra* corn enough to keep them for a year. This is probably a large estimate, but it serves to show that this fertilizer is very valuable, and also that poultry may be kept with great profit, if their excrements are properly secured.

The manure of pigeons has been a favorite fertilizer in some countries for more than 2000 years.

Market gardeners attach much value to rabbit-manure.

SHEEP MANURE.

The manure of sheep is less valuable than it would be, if so large a quantity of the nitrogen and mineral parts of the food were not employed in the formation of wool. This has a great effect on the richness of the excrements, but they are still a very good fertilizer, and should be protected from loss in the same way as stable manure.

What can you say of the manure of sheep?

GUANO.

Guano as a manure has become world renowned. The worn-out tobacco lands of Virginia, and other fields in many parts of the country, which seemed to have yielded to the effect of an ignorant course of cultivation, and to have sunk to their final repose, have in many cases been revived to the production of excellent crops, and have had their value multiplied many fold by the use of guano. Although an excellent manure, it should not cause us to lose sight of those valuable materials which exist on almost every farm. Every ton of guano imported into the United States is an addition to our national wealth, but every ton of stable-manure, or poultry-dung, or night soil evaporated or carried away in rivers, is equally a *deduction* from our riches. If the imported manure is to really benefit us, we must not allow it to occasion the neglect and consequent loss of our domestic fertilizers.

The Peruvian guano (which is considered the best) is brought from islands near the coast of Peru. The birds which frequent these islands live almost entirely on fish, and drop their excrements here in a climate where rain is almost unknown, and where, from the dryness of the air, there is but little loss

---

Should the use of guano induce us to disregard other manures?
Where and in what manner is the best guano deposited?

sustained by the manure. It is brought to this country in large quantities, and is an excellent fertilizer, superior even to night soil.

It should be mixed with an absorbent before being used, unless it is plowed deeply under the soil, as it contains much ammonia which would be lost from evaporation. It would probably also injure plants. The best way to use guano, is in connection with sulphuric acid and bones, as will be described hereafter.

The composition of the various kinds of guano may be found in the section on analysis.

CHAPTER VII.

OTHER ORGANIC MANURES.

THE number of organic manures is almost countless. The most common of these have been described in the previous chapters on the excrements of animals. The more prominent of the remaining ones will now be considered. As a universal rule, it may be stated that all organic matter (every thing which has had vegetable or animal life) is capable of fertilizing plants.

How should it be prepared for use?

DEAD ANIMALS.

The bodies of animals contain much *nitrogen*, as well as valuable quantities, the phosphates and other inorganic materials required in the growth of plants. On their decay, the nitrogen is resolved into *ammonia*,* and the mineral matters become valuable as food for the inorganic parts of plants.

If the decomposition of animal bodies takes place in exposed situations, and without proper precautions, the ammonia escapes into the atmosphere, and much of the mineral portion is leached out by rains. The use of absorbents, such as charcoal-dust, prepared muck, etc., will entirely prevent evaporation, and will in a great measure serve as a protection against leaching.

If a dead horse be cut in pieces and mixed with ten loads of muck, the whole mass will, in a single season, become a most valuable compost. Small animals, such as dogs, cats, etc., may be with advantage buried by the roots of grape-vines or trees.

* Under some circumstances, *nitric acid* is formed, which is equally beneficial to vegetable growth.

What are the chief fertilizing constituents of dead animals?
What becomes of these when exposed to the atmosphere?
How may this be prevented?

BONES.

The *bones* of animals contain phosphate of lime and gelatine. The gelatine is a nitrogenous substance, and produces ammonia on its decomposition. This subject will be spoken of more fully under the head of 'phosphate of lime' in the chapter on mineral manures, as the treatment of bones is more directly with reference to the fertilizing value of their inorganic matter.

FISH.

In many localities near the sea-shore large quantities of fish are caught and applied to the soil. These make excellent manure. They contain much nitrogen, which renders them strongly ammoniacal on decomposition. Their bones consist of phosphate and carbonate of lime; and, being naturally soft, they decompose in the soil with great facility, and become available to plants. The scales of fish contain valuable quantities of nitrogen, phosphate of lime, etc., all of which are highly useful.

Refuse fishy matters from markets and from the house are well worth saving. These and fish caught for manure may be made into compost with prepared

---

Of what do the bones of animals consist?
What is gelatine?
Describe the fertilizing qualities of fish.

muck, etc.; and, as they putrefy rapidly, they soon become ready for use. They may be added to the compost of stable manure with great advantage.

Fish (like all other nitrogenous manures) should never be applied as a top dressing, unless previously mixed with a good absorbent of ammonia, but should when used alone be immediately plowed under to considerable depth, to prevent the evaporation—and consequent loss—of their fertilizing gases.

WOOLLEN RAGS, ETC.

Woollen rags, hair, waste of woollen factories, etc., contain both nitrogen and phosphate of lime; and, like all other matters containing these ingredients, are excellent manures, but must be used in such a way as to prevent the escape of their fertilizing gases. They decompose slowly, and are therefore considered a *lasting* manure. Like all *lasting* manures, however, they are *slow* in their effects, and the most advantageous way to use them is to compost them with stable manure, or with some other rapidly fermenting substance, which will hasten their decomposition and render them sooner available.

Rags, hair, etc., thus treated, will in a short time be reduced to such a condition that they may be immediately used by plants instead of lying in the

---

Should these be applied as a top dressing to the soil?
What are the fertilizing properties of woollen rags?
What is the best way to use them?

soil to be slowly taken up. It is better in all cases to have manures act *quickly* and give an immediate return for their cost, than to lie for a long time in the soil before their influence is felt.

A pound of woollen rags is worth, as a manure, twice as much as is paid for good linen shreds for paper making ; still, while the latter are always preserved, the former are thrown away, although considered by good judges to be worth forty times as much as barn-yard manure.

Old leather should not be thrown away. It decomposes very slowly, and consequently is of but a little value ; but, if put at the roots of young trees, it will in time produce appreciable effects.

Tanners' and curriers' refuse, and all other animal offal, including that of the slaughter-house, is well worth attention, as it contains more or less of those two most important ingredients of manures, nitrogen and phosphate of lime.

It is unnecessary to add that, in common with all other animal manures, these substances must be either composted, or immediately plowed under the soil. Horn piths, and horn shavings, if decomposed in compost, with substances which ferment rapidly, make very good manure, and are worth fully the price charged for them.

What is their value compared with that of farm-yard manure?
How should old leather be treated?
Describe the manurial properties of tanners' refuse.
How should they be treated?
Are horn piths, etc. valuable?

ORGANIC MANURES OF VEGETABLE ORIGIN.

Muck, the most important of the purely vegetable manures, has been already sufficiently described. It should be particularly borne in mind that, when first taken from the swamp it is often *sour*, or *cold*, but that if exposed for a long time to the air, or if well treated with lime, unleached ashes, the lime and salt mixture, or any other alkali, its acids will be *neutralized* (or overcome), and it becomes a good application to any soil, except peat or other soils already containing large quantities of organic matter. In applying muck to the soil (as has been before stated), it should be made a vehicle for carrying ammoniacal manures.

SPENT TAN BARK.

Spent tan bark, if previously decomposed by the use of the lime and salt mixture, or potash, answers all the purposes of prepared muck, but is more difficult of decomposition.

The bark of trees contains a larger proportion of inorganic matter than the wood, and much of this, on the decomposition of the bark, becomes available as manure. The chemical effect on the bark, of

---

Why is decomposed bark more fertilizing than that of decayed wood?

using it in the tanning of leather, is such as to render it difficult to be rotted by the ordinary means, but, by the use of the lime and salt mixture it may be reduced to the finest condition, and becomes a most excellent manure. It probably contains small quantities of nitrogen (obtained from the leather), which adds to its value. Unless tan bark be composted with lime, or some other alkali, it may produce injurious effects from the *tannic acid* which it is liable to contain. Alkaline substances will neutralize this acid, and prevent it from being injurious.

One great benefit resulting from the use of spent tan bark, is due to its power of absorbing moisture from the atmosphere. For this reason it is very valuable for *mulching** young trees and plants when first set out.

SAWDUST.

Sawdust in its natural state is of very little value to the land, but when decomposed, as may be done by the same method as was described for tan bark, it is of some importance, as it contains a large quantity of carbon. Its ash, too, which becomes

* See the glossary at the end of the book.

How may bark be decomposed?
Why should tan bark be composted with an alkali?
Why is it good for mulching?
Is sawdust of any value?

available, contains soluble inorganic matter, and in this way it acts as a direct manure. So far as concerns the value of the ash, however, the bark is superior to sawdust. Sawdust may be partially rotted by mixing it with strong manure (as hog manure), while it acts as a *divisor*, and prevents the too rapid action of this when applied to the soil. Some kinds of sawdust, such as that from beech wood, form acetic acid on their decomposition, and these should be treated with, at least, a sufficient quantity of lime to correct the acid.

Soot is a good manure. It contains much carbon, and has, thus far, all of the beneficial effects of charcoal dust. The sulphur, which is one of its constituents, not only serves as food for plants, but, from its odor, is a good protection against some insects. By throwing a handful of soot on a melon vine, or young cabbage plant, it will keep away many insects.

Soot contains some ammonia, and as this is in the form of a *sulphate*, it is not volatile, and consequently does not evaporate when the soot is applied as a top dressing, which is the almost universal custom.

Why is sawdust a good addition to the pigstye?
What is the peculiarity of sawdust from the beech, etc.?
What is a peculiarity of soot?
Why may soot be used as a top dressing without losing its ammonia?

GREEN CROPS.

Green crops, to plough under, are in many places largely raised, and are always beneficial. The plants most used for this purpose, in our country, are clover, buckwheat, and peas. These plants have very long roots, which they send deep in the soil, to draw up mineral matter for their support. This mineral matter is deposited in the plant. The leaves and roots receive carbonic acid and ammonia from the air, and from water. In this manner they obtain their carbon. When the crop is turned under the soil, it decomposes, and the carbon, as well as the mineral ingredients obtained from the subsoil, are deposited in the surface soil, and become of use to succeeding crops. The hollow stalks of the buckwheat and pea, serve as tubes, in the soil, for the passage of air, and thus, in heavy soils, give a much needed circulation of atmospheric fertilizers.

Although green crops are of great benefit, and are managed with little labor, there is no doubt but the same results may be more economically produced. A few loads of prepared muck will do more towards increasing the organic matter in the soil, than a very heavy crop of clover, while it would be ready for immediate cultivation, instead of having to lie idle

---

What plants are most used as green crops?
What office is performed by the roots of green crops?
How do such manures increase the organic matter of soils?

during the year required in the production and decomposition of the green crop. The effect of the roots penetrating the subsoil is, as we have seen, to draw up inorganic matter, to be deposited within reach of the roots of future crops. In the next section we shall show that this end may be much more efficiently attained by the use of the subsoil plow, which makes a passage for the roots into the subsoil, where they can obtain for themselves what would, in the other case, be brought up for them by the roots of the green crop.

The offices of the hollow straws may be performed by a system of ridging and back furrowing, having previously covered the soil with leaves, or other refuse organic material.

In *high farming*, where the object of the cultivator is to make a profitable investment of labor, these last named methods will be found most expedient ; but, if the farmer have a large quantity of land, and can afford but a limited amount of labor, the raising of green crops, to be plowed under in the fall, will probably be adopted.

Before closing this chapter, it may be well to remark that there are various other fertilizers, such as

---

What office is performed by the straw of the buckwheat and pea?

What treatment may be substituted for the use of green crops?

Which course should be adopted in high farming?

Why is the use of green crops preferable in ordinary cultivation?

Name some other valuable manures.

the *ammoniacal liquor of gas-houses, soapers' wastes, bleachers' lye, lees of old oil casks, etc.*, which we have not space to consider at length, but which are all valuable as additions to the compost heap, or as applications, in a liquid form, to the soil.

In many cases (when heavy manuring is practised), it may be well to apply organic manures to the soil in a green state, turn them under, and allow them to undergo decomposition in the ground. The advantages of this system are, that the *heat*, resulting from the chemical changes, will hasten the growth of plants, by making the soil warmer; the carbonic acid formed will be presented to the roots instead of escaping into the atmosphere; and if the soil be heavy, the rising of the gases will tend to loosen it, and the leaving vacant of the spaces occupied by the solid matters will, on their being resolved into gases, render the soil of a more porous character. As a general rule, however, in ordinary farming, where the amount of manure applied is only sufficient for the supply of food to the crop, it is undoubtedly better to have it previously decomposed—*cooked* as it were, for the uses of the plants—as they can then obtain the required amount of nutriment as fast as needed.

What are the advantages arising from burying manure in its green state?

Which is generally preferable, this course, or composting? Why?

ABSORPTION OF MOISTURE.

It is often convenient to know the relative power of different manures to absorb moisture from the atmosphere, especially when we wish to manure lands that suffer from drought. The following results are given by C. W. Johnson, in his essay on salt, (pp. 8 and 19). In these experiments the animal manures were employed without any admixture of straw.

		PARTS
1000 parts	of horse dung, dried in a temperature of 100°, absorbed by exposure for three hours, to air saturated with moisture, of the temperature of 62°	145
1000 parts	of cow dung, under the same circumstances, absorbed	130
1000 parts	pig dung	120
1000 "	sheep "	81
1000 "	pigeon "	50
1000 "	rich alluvial soil	14
1000 "	fresh tanner's bark	115
1000 "	putrified "	145
1000 "	refuse marine salt sold as manure	49½
1000 "	soot	36
1000 "	burnt clay	29
1000 "	coal ashes	14
1000 "	lime	11

			PARTS.
1000	parts	sediment from salt pans	10
1000	"	crushed rock salt	10
1000	"	gypsum	9
1000	"	salt	4*

Muck is a most excellent absorbent of moisture, when thoroughly decomposed.

DISTRIBUTION OF MANURES.

The following table from Johnson, on manures, will be found convenient in the distribution of mannures.

By its assistance the farmer will know how many loads of manure he requires, dividing each load into a stated number of heaps, and placing them at certain distances. In this manner manure may be applied evenly, and calculation may be made as to the amount, per acre, which a certain quantity will supply.†

* Working Farmer, vol. 1, p. 55.

† It is not necessary that this and the foregoing table should be learned by the scholar, but they will be found valuable for reference by the farmer.

DISTANCE OF THE HEAPS.	NUMBER OF HEAPS IN A LOAD.									
	1	2	3	4	5	6	7	8	9	10
3 yards. . .	538	269	179	134	108	89½	77	67	60	54
3½ do. . .	395	168	132	99	79	66	56½	49½	44	39½
4 do. . .	303	151	101	75½	60½	50½	43¼	37¾	33½	30¼
4½ do. . .	239	120	79½	60	47¾	39¾	34¼	30	26½	24
5 do. . .	194	97	64½	48½	38¾	32¼	27¾	24¼	21½	19¼
5½ do. . .	160	80	53½	40	32	26¾	22¾	20	17¾	16
6 do. . .	131	67	44¾	33½	27	22½	19¼	16¾	15	13½
6½ do. . .	115	57½	38¼	28¾	23	19	16¼	14¼	12¾	11½
7 do. . .	99	49½	33	24¾	19¾	16½	14	12¼	11	10
7½ do. . .	86	43	28¾	21½	17¼	14¼	12¼	10¾	9½	8½
8 do. . .	75½	37¾	25¼	19	15¾	12½	10¾	9½	8½	7½
8½ do. . .	67	33½	22¼	16¾	13½	11¼	9½	8½	7½	6¾
9 do. . .	60	30	20	15	12	10	8½	7¾	6¾	6
9½ do. . .	53½	26¾	18	13½	10¾	9	7¾	6¾	6	5¼
10 do. . .	48½	24¼	16¼	12	9¾	8	7	6	5½	4¾

Example 1.—Required, the number of loads necessary to manure an acre of ground, dividing each load into six heaps, and placing them at a distance of 4½ yards from each other? The answer by the table is 39¾.

Example 2.—A farmer has a field containing 5½ acres, over which he wishes to spread 82 loads of dung. Now 82 divided by 5½, gives 15 loads per acre; and by referring to the table, it will be seen that the desired object may be accomplished, by making 4 heaps of a load, and placing them 9 yards apart, or by 9 heaps at 6 yards, as may be thought advisable.

CHAPTER VIII.

MINERAL MANURES.

THE second class of manures named in the general division of the subject, in the early part of this chapter, comprises those of a mineral character, or *inorganic* manures.

These manures have four kinds of action when applied to the soil.

1st. They furnish food for the inorganic part of plants.

2d. They prepare matters already in the soil, for assimilation by roots.

3d. They improve the mechanical condition of the soil.

4th. They absorb ammonia.

Some of the mineral manures produce in the soil only one of these effects, and others are efficient in two or all of them.

The principles to be considered in the use of mineral manures are essentially given in the first two sections of this book. It may be well, however, to repeat them briefly in this connection, and to give the *reasons* why any of these manures are needed, from which we may learn what rules are to be observed in their application.

1st. Those which are used as food by plants. It will be recollected that the *ash* left after burning plants, and which formed a part of their structures, has a certain chemical composition; that is, it consists of alkalies, acids, and neutrals. It was also

---

How many kinds of action have inorganic manures? What is the first of these? The second? Third? Fourth? Do all mineral manures possess all of these qualities?

stated that the ashes of plants of the same kind are always of about the same composition, while the ashes of different kinds of plants may vary materially. Different parts of the same plant too, as we learned, are supplied with different kinds of ash.

For instance, *clover*, on being burned, leaves an ash containing *lime*, as one of its principal ingredients, while the ash of *potatoes* contains more of *potash* than of any thing else.

In the second section (on soils), we learned that some soils contain every thing necessary to make the ashes of all plants, and in sufficient quantity to supply what is required, while other soils are either entirely deficient in one or more ingredients, or contain so little of them that they are unfertile for certain plants.

From this, we see that we may pursue either one of two courses. After we know the exact composition of the soil—which we can learn only from correct analysis—we may manure it with a view either to making it fertile for all kinds of plants or only for one particular plant. For instance, we may find that a soil contains a very little phosphoric acid, and no potash. If we wish to raise potatoes on such a soil, we have only to apply potash (if the soil is good

---

Relate what you know of the properties of vegetable ashes?
How does this relate to the fertility of the soil?
According to what two rules may we apply mineral manures?
What course would you pursue to raise potatoes on a soil containing a very little phosphoric acid and no potash?

in other particulars), which is largely required by this plant, though it needs but little phosphoric acid; while, if we wish to make it fertile for wheat, and all other plants, we must apply more phosphoric acid as well as postash. As a universal rule, it may be stated that to render a soil fertile for any particular plant, we must supply it (unless it already contains them) with those matters which are necessary to *make* the ash of that plant; and, if we would render it capable of producing *all* kinds of plants, it must be furnished with the materials required in the formation of *all kinds of vegetable ashes.*

It is not absolutely necessary to have the soil analyzed before it can be cultivated with success, but it is the *cheapest* way.

We might proceed from an analysis of the plant required (which will be found in Section V.), and apply to the soil in the form of manure every thing that is necessary for the formation of the ash of that plant. This would give a good crop on *any* soil that was in the proper *mechanical* condition, and contained enough organic matter; but a moment's reflection will show that, if the soil contained a large amount of potash, or of phosphate of lime, it would not be necessary to make an application of more of these ingredients—at an expense of perhaps three times the cost of an analysis. It is true that, if the

---

Would you manure it in the same way for wheat?
Why?

crop is *sold*, and it is desired to maintain the fertility of the soil, the full amount of the ash must be applied, either before or after the crop is grown ; but, in the ordinary use of crops for feeding purposes, a large part of the ash will exist in the excrements of the animals ; so that the judicious farmer will be able to manure his land with more economy than if he had to apply to each crop the whole amount and variety required for its ash. The best rule for practical manuring is probably to *strengthen the soil in its weaker points, and prevent the stronger ones from becoming weaker.* In this way, the soil may be raised to the highest state of fertility, and be fully maintained in its productive powers.

2d. Those manures which render available matter already contained in the soil.

Silica (or sand), it will be recollected, exists in all soils ; but, in its pure state, is not capable of being dissolved, and therefore cannot be used by plants. The alkalies (as has been stated), have the power of combining with this silica, making compounds, which are called *silicates*. These are readily dissolved by water, and are available in vegetable growth. Now, if a soil is deficient in these soluble silicates, it is well known that grain, etc.,

---

How is the fertility of the soil to be maintained, if the crops are *sold?*

What rule is given for general treatment?

Give an instance of matters in the soil that are to be rendered available by mineral manures?

grown on it, not being able to obtain the material which gives them strength, will fall down or *lodge*; but, if such measures be taken, as will render the sand soluble, the straw will be strong and healthy. Alkalies used for this purpose, come under the head of those manures which develope the natural resources of the soil.

Again, much of the mineral matter in the soil is combined within particles, and is therefore out of the reach of roots. Lime, among other thing, has the effect of causing these particles to crumble and expose their constituents to the demand of roots. Therefore, lime has for one of its offices the development of the fertilizing ingredients of the soil.

3d. Those manures which improve the mechanical condition of the soil.

The alkalies, in combining with sand, commence their action on the surfaces of the particles, and roughen them—*rust* them as it were. This roughening of particles of the soil prevents them from moving among each other as easily as they do when they are smooth, and thus keeps the soil from being compacted by heavy rains, as it is liable to be in its natural condition. In this way, the mechanical texture of the soil is improved.

It has just been said that *lime* causes the pul-

How may silica be developed?
How does lime affect soils containing coarse particles?
How do mineral manures sometimes improve the mechanical texture of the soil?

verization of the particles of the soil; and thus, by making it finer, improves its mechanical condition.

Some mineral manures, as plaster and salt, have the power of absorbing moisture from the atmosphere; and this is a mechanical improvement to dry soils.

4th. Those mineral manures which have the power of absorbing ammonia.

Plaster, *chloride of lime*, *alumina* (*clay*), etc., are large absorbents of ammonia, whether arising from the fermentation of animal manures or washed down from the atmosphere by rains. The ammonia thus absorbed is of course very important in the vegetation of crops.

Having now explained the reasons why mineral manures are necessary, and the manner in which they produce their effects, we will proceed to examine the various deficiencies of soils and the character of many kinds of this class of fertilizers.

CHAPTER IX.

DEFICIENCIES OF SOILS, MEANS OF RESTORATION, ETC.

As will be seen by referring to the analyses of soils

Name some mineral manures which absorb ammonia?

on p. 72, they may be deficient in certain ingredients, which it is the object of mineral manures to supply. These we will take up in order, and endeavor to show in a simple manner the best means of managing them in practical farming.

ALKALIES.

POTASH.

Potash is often deficient in the soil. Its deficiency may have been caused in two ways. Either it may not have existed largely in the rock from which the soil was formed, and consequently is equally absent from the soil itself, or it may have once been present in sufficient quantities, and been carried away in crops, without being returned to the soil in the form of manure until too little remains for the requirements of fertility.

In either case, its absence may be accurately detected by a skilful chemist, and it may be supplied by the farmer in various ways. Potash, as well as all of the other mineral manures, is contained in the excrements of animals, but not (as is also the case with the others) in sufficient quantities to restore the proper balance to soils where it is largely

---

Do all soils contain a sufficient amount of potash?
How may its deficiency have been caused?
How may its absence be detected?
Does barn-yard manure contain sufficient potash to supply its deficiency in worn-out soils?

deficient, nor even to make up for what is yearly removed with each crop, except that crop (or its equivalent) has been fed to such animals as return *all* of the fertilizing constituents of their food in the form of manure, and this be all carefully preserved and applied to the soil. In all other cases, it is necessary to apply more potash than is contained in the excrements of animals.

Unleached wood ashes is generally the most available source from which to obtain this alkali. The ashes of all kinds of wood contain potash (more or less according to the kind—see analysis section V.) If the ashes are *leached,* the potash is removed; and, hence for the purpose of suplying it, they are worthless; but *unleached* ashes are an excellent source from which to obtain it. They may be made into compost with muck, as directed in a previous chapter, or applied directly to the soil. In either case the potash is available directly to the plant, or is capable of uniting with the silica in the soil to form silicate of potash. Neither potash nor any other alkali should ever be applied to animal manures unless in compost with an absorbent, as they cause the ammonia to be thrown off and lost.

Potash sparlings, or the refuse of potash ware-

What is generally the most available source from which to obtain this alkali?
Will leached ashes answer the same purpose?
How may ashes be used?

houses, is an excellent manure for lands deficient in this constituent.

Potash marl, such as is found in New Jersey, contains a large proportion of potash, and is an excellent application to soils requiring it.

Feldspar, kaolin, and other minerals containing potash, are, in some localities, to be obtained in sufficient quantities to be used for manurial purposes.

Granite contains potash, and if it can be crushed (as is the case with some of the softer kinds,) it serves a very good purpose.

SODA.

Soda, the requirement of which is occasioned by the same causes as create a deficiency of potash, and all of the other ingredients of vegetable ashes, may be very readily supplied by the use of *common salt* (chloride of sodium), which consists of about one half sodium (the base of soda). The best way to use salt is in the lime and salt mixture, previously described, or as a direct application to the soil. If too much salt be given to the soil it will kill any plant. In small quantities, however, it is highly beneficial, and if *six bushels per acre* be sown broadcast over the land, to be carried in by rains and dews,

---

From what other sources may potash be obtained?
How may we obtain soda?
In what quantities should pure salt be applied to the soil?

it will not only destroy many insects (grubs, worms, etc.), but will, after decomposing and becoming chlorine and soda, prove an excellent manure. Salt, even in quantities large enough to denude the soil of all vegetation, is never *permanently* injurious. After the first year, it becomes resolved into its constituents, and furnishes chlorine and soda to plants, without injuring them. One bushel of salt in each cord of compost will not only hasten the decomposition of the manures, but will kill all seeds and grubs—a very desirable effect. While small quantities of salt in a compost heap are beneficial, too much (as when applied to the soil) is positively injurious, as it arrests decomposition; fairly *pickles* the manures, and prevents them from rotting.

For *asparagus*, which is a marine plant, salt is an excellent manure, and may be applied in almost unlimited quantities, *while the plants are growing*, if used after they have gone to top, it is injurious. Salt has been applied to asparagus beds in such quantities as to completely cover them, and with apparent benefit to the plants. Of course large doses of salt kill all weeds, and thus save labor and the injury to the asparagus roots, which would result from their removal by hoeing. Salt may be used advantageously in any of the foregoing manners, but should always be applied with care. For ordinary farm purposes,

---

If applied in large quantities will it produce permanent injury? In what quantities should salt be applied to composts? To asparagus?

it is undoubtedly most profitable to use the salt with lime, and make it perform the double duty of assisting in the decomposition of vegetable matter, and fertilizing the soil.

Soda unites with the silica in the soil, and forms the valuable *silicate of soda.*

Nitrate of soda, or cubical nitre, which is found in South America, consists of soda and nitric acid. It furnishes both soda and nitrogen to plants, and is an excellent manure.

LIME.

The subject of *lime* is one of most vital importance to the farmer; indeed, so varied are its modes of action and its effects, that some writers have given it credit for every thing good in the way of farming, and have gone so far as to say that *all* permanent improvement of agriculture must depend on the use of lime. Although this is far in excess of the truth (as lime cannot plow, nor drain, nor supply any thing but *lime* to the soil), its many beneficial effects demand for it the closest attention.

As food for plants, lime is of considerable importance. All plants contain lime—some of them in large quantities. It is an important constituent of

---

What is generally the best way to use salt?
What is nitrate of soda?
What plants contain lime?

straw, meadow hay, leaves of fruit trees, peas, beans, and turnips. It constitutes more than one third of the ash of red clover. Many soils contain lime enough for the use of plants, in others it is deficient, and must be supplied artificially before they can produce good crops of those plants of which lime is an important ingredient. The only way in which the exact quantity of lime in the soil can be ascertained is by chemical analysis. However, the amount required for the mere feeding plants is not large, (much less than one per cent.), but lime is often necessary for other purposes ; and setting aside, for the present, its feeding action, we will examine its various effects on the mechanical and chemical condition of the soil.

1. It corrects acidity (sourness).

2. It hastens the decomposition of the organic matter in the soil.

3. It causes the mineral particles of the soil to crumble.

4. By producing the above effects, it prepares the constituents of the soil for assimilation by plants.

5. It is *said* to exhaust the soil, but it does so in a very desirable manner, the injurious effects of which may be easily avoided.

1. The decomposition of organic matter in the

Do all soils contain enough lime for the use of plants?
What amount is needed for this purpose?
What is its first-named effect on the soil?
Its second? Third? Fourth? Fifth?
How are acids produced in the soil?

soil, often produces acids which makes the land *sour*, and cause it to produce sorrel and other weeds, which interfere with the healthy growth of crops. Lime is an *alkali*, and if applied to soils suffering from sourness, it will unite with the acids, and neutralize them, so that they will no longer be injurious.

2. We have before stated that lime is a decomposing agent, and hastens the rotting of muck and other organic matter. It has the same effect on the organic parts of the soil, and causes them to be resolved into the gases and minerals of which they are formed. It has this effect, especially, on organic matters containing *nitrogen*, causing them to throw off ammonia ; consequently, it liberates this gas from the animal manures in the soil.

3. Various inorganic compounds in the soil are so affected by lime, that they lose their power of holding together, and crumble, or are reduced to finer particles, while some of their constituents are rendered soluble. One way in which this is accomplished is by the action of the lime on the silica contained in these compounds, forming the silicate of lime. This crumbling effect improves the mechanical as well as the chemical condition of the soil.

4. We are now enabled to see how lime prepares the constituents of the soil for the use of plants.

How does lime correct them?
How does it affect animal manures in the soil?

By its action on the roots, buried stubble, and other organic matter in the soil, it causes them to be decomposed, and to give up many of their gaseous and inorganic constituents for the use of roots. In this manner the organic matter is prepared for use more rapidly than would be the case, if there were no lime present to hasten its decomposition.

By the decomposing action of lime on the mineral parts of the soil (3), they also are placed more rapidly in a useful condition than would be the case, if their preparation depended on the slow action of atmospheric influences.

Thus, we see that lime, aside from its use directly as food for plants, exerts a beneficial influence on both the organic and inorganic parts of the soil.

5. Many contend that lime *exhausts* the soil.

If we examine the manner in which it does so, we shall see that this is no argument against its use.

It exhausts the organic parts of the soil, by decomposing them, and resolving them into the gases and minerals of which they are composed. If the soil do not contain a sufficient quantity of absorbent matter, such as clay or charcoal, the gases arising from the organic matter are liable to escape; but when there is a sufficient amount of these substances present (as there always should be), these gases are

---

Inorganic compounds?
How does lime prepare the constituents of the soil for use?
What can you say of the remark that lime exhausts the organic matter in the soil?

all retained until required by the roots of plants. Hence, although the organic matter of manure and vegetable substances may be *altered in form*, by the use of lime, it can escape (except in very poor soils) only as it is taken up by roots to feed the crop, and such exhaustion is certainly profitable; still, in order that the fertility of the soil may be *maintained*, enough of organic manure should be applied, to make up for the amount taken from the soil by the crop, after liberation for its use by the action of the lime. This will be but a small proportion of the organic matter contained in the crop, as it obtains the larger part from the atmosphere.

The only way in which lime can exhaust the inorganic part of the soil is, by altering its condition, so that plants can use it more readily. That is, it exposes it for solution in water. We have seen that fertilizing matter cannot be leached out of a good soil, in any material quantity, but can only be carried down to a depth of about thirty-four inches. Hence, we see that there can be no loss in this direction; and, as inorganic matter cannot evaporate from the soil, the only way in which it can escape is through the structure of plants.

If lime is applied to the soil, and increases the amount of crops grown by furnishing a larger supply of inorganic matter, of course, the removal of inor-

---

How can lime exhaust the mineral parts of the soil?
Must the matter taken away be returned to the soil?

ganic substances from the soil will be more rapid than when only a small amount of crop is grown, and the soil will be sooner exhausted—not by the lime, but by the plants. In order to make up for this exhaustion, it is necessary that a sufficient amount of inorganic matter be supplied to compensate for the increased quantity taken away by plants.

Thus we see, that it is hardly fair to accuse the *lime* of exhausting the soil, when it only improves its character, and increases the amount of its yield. It is the *crop* that takes away the fertility of the soil (the same as would be the case if no lime were used, only faster as the crop is larger), and in all judicious cultivation, this loss will be fully compensated by the application of manures, thereby preventing the exhaustion of the soil.

Kind of lime to be used. The first consideration in procuring lime for manuring land, is to select that which contains but little, if any *magnesia.* Nearly all stone lime contains more or less of this, but some kinds contain more than others. When magnesia is applied to the soil, in too large quantities, it is positively injurious to plants, and great care is necessary in making selection. As a general rule, it may

---

If this course be pursued, will the soil suffer from the use of lime?
Is it the lime, or its crop, that exhausts the soil?
Is lime containing magnesia better than pure lime?
What is the best kind of lime?

be stated, that the best plastering lime makes the best manure. Such kinds only should be used as are known from experiment not to be injurious.

Shell lime is undoubtedly the best of all, for it contains no magnesia, and it does contain a small quantity of *phosphate of lime.* In the vicinity of the sea-coast, and near the lines of railroads, oyster shells, clam shells, etc., can be cheaply procured. These may be prepared for use in the same manner as stone lime.*

The preparation of the lime is done by first burning and then slaking, or by putting it directly on the land, in an unslaked condition, after its having been burned. Shells are sometimes *ground,* and used without burning; this is hardly advisable, as they cannot be made so fine as by burning and slaking. As was stated in the first section of this book, lime usually exists in nature, in the form of carbonate of lime, as limestone, chalk, or marble (being lime and carbonic acid combined), and when this is burned, the carbonic acid is thrown off, leaving the lime in a pure or caustic form. This is called burned lime, quick-lime, lime shells, hot lime, etc. If

* Marl is earth containing lime, but its use is not to be recommended in this country, except where it can be obtained at little cost, as the expenses of carting the *earth* would often be more than the value of the *lime.*

Is the purchase of marl to be recommended?
How is lime prepared for use? (Note.)
Describe the burning and slaking of lime.

the proper quantity of water be poured on it, it is immediately taken up by the lime, which falls into a dry powder, called *slaked lime.* If *quick-lime* were left exposed to the weather, it would absorb moisture from the atmosphere, and become what is termed *air slaked.*

When *slaked lime* (consisting of lime and water) is exposed to the atmosphere, it absorbs carbonic acid, and becomes carbonate of lime again; but it is now in the form of a very fine powder, and is much more useful than when in the stone.

If quick-lime is applied directly to the soil, it absorbs first moisture, and then carbonic acid, becoming finally a powdered carbonate of lime.

One ton of *carbonate of lime* contains 11¼ cwt. of lime; the remainder is carbonic acid. One ton of *slaked lime* contains about 15 cwt. of lime; the remainder is water.

Hence we see that lime should be burned, and not slaked, before being transported, as it would be unprofitable to transport the large quantity of carbonia acid and water contained in carbonate of lime and slaked lime. The quick-lime may be slaked,

---

What is air slaking?

If slaked lime be exposed to the air, what change does it undergo?

What is the object of slaking lime?

How much carbonic acid is contained in a ton of carbonate of lime?

How much lime does a ton of slaked lime contain?

What is the most economical form for transportation?

and carbonated after reaching its destination, either before or after being applied to the land.

As has been before stated, much is gained by slaking lime with *salt water*, thus imitating the lime and salt mixture. Indeed in many cases, it will be found profitable to use all lime in this way. Where a direct action on the inorganic matters contained in the soil is desired, it may be well to apply the lime directly in the form of quick-lime; but, where the decomposition of the vegetable and animal constituents of the soil is desired, the correction of *sourness*, or the supplying of lime to the crop, the mixture with salt would be advisable.

The amount of lime required *by plants* is, as was before observed, usually small compared with the whole amount contained in the soil; still it is not unimportant.

				OF LIME.	
25	bus. of	wheat	contain about	13	lbs.
25	"	barley	"	10½	"
25	"	oats	"	11	"
2	tons of	turnips	"	12	"
2	"	potatoes	"	5	"
2	"	red clover	"	77	"
2	"	rye grass	"	30	" *

* The straw producing the grain and the turnip and potato tops contain more lime than the grain and roots.

What is the best form for immediate action on the inorganic matter in the soil?

For most other purposes?

The amount of lime required at each application, and the frequency of those applications, must depend on the chemical and mechanical condition of the soil. No exact rule can be given, but probably the custom of each district—regulated by long experience—is the best guide.

Lime sinks in the soil; and therefore, when used alone, should always be applied as a top dressing to be carried into the soil by rains. The tendency of lime to settle is so great that, when cutting drains, it may often be observed in a whitish streak on the top of the subsoil. After heavy doses of lime have been given to the soil, and have settled so as to have apparently ceased from their action, they may be brought up and mixed with the soil by deeper plowing.

Lime should never be mixed with animal manures, unless in compost with muck, or some other good absorbent, as it is liable to cause the escape of their ammonia.

PLASTER OF PARIS.

Plaster of Paris or Gypsum (sulphate of lime) is composed of sulphuric acid and lime in combination. It is called 'plaster of Paris,' because it constitutes the rock underlying the city of Paris.

What is the best guide concerning the quantity of lime to be applied?
What is said of the sinking of lime in the soil?
What is plaster of Paris composed of?
Why is it called plaster of Paris?

It is a constituent of many plants. It also furnishes them with sulphur—a constituent of the sulphuric acid which it contains.

It is an excellent absorbent of ammonia, and is very useful to sprinkle around stables, poultry houses, pig-styes, and privies, where it absorbs the escaping gases, saving them for the use of plants, and purifying the air, thus rendering stables, etc., more healthy than when not so supplied.

It has been observed that the extravagant use of plaster sometimes induces the growth of *sorrel*. This is probably the case only where the soil is deficient in lime. In such instances, the lime required by plants is obtained by the decomposition of the plaster. The lime enters into the construction of the plant, and the sulphuric acid remains *free*, rendering the soil *sour*, and therefore in condition to produce sorrel. In such a case, an application of *lime* will correct the acid by uniting with it and converting it into *plaster*.

CHLORIDE OF LIME.

Chloride of lime is a compound of *lime and chlorine*. It furnishes both of these constituents to plants, and it is an excellent absorbent of ammonia

---

Is it a constituent of plants?
What else does it furnish them?
How does it affect manure?
How does it produce sorrel in the soil?
How may the acidity be overcome?

and other gases arising from decomposition—hence its usefulness in destroying bad odors, and in preserving fertilizing matters for the use of crops.

It may be used like plaster, or in the decomposition of organic matters, where it not only hastens decay, but absorbs and retains the escaping gases. It will be recollected that *chloride of lime* is one of the products of the *lime and salt mixture.*

Lime in combination with *phosphoric acid* forms the valuable *phosphate of lime,* of which so large a portion of the ash of grain, and the bones of animals, is formed. This will be spoken of more at length under the head of 'phosphoric acid.'

MAGNESIA.

Magnesia is a constituent of vegetable ashes, and is almost always present in the soil in sufficient quantities. When analysis indicates that it is needed, it may be applied in the form of *magnesian lime,* or *refuse epsom salts,* which are composed of sulphuric acid and magnesia (sulphate of magnesia).

The great care necessary concerning the use of magnesia is, not to apply too much of it, it being,

---

What does chloride of lime supply to plants?
How does it affect manures?
How may it be used?
How may magnesia be supplied, when wanting?
What care is necessary concerning the use of magnesia?

when in excess, as has been previously remarked, injurious to the fertility of the soil. Some soils are hopelessly barren from the fact that they contain too much magnesia.

ACIDS.

SULPHURIC ACID.

Sulphuric acid is a very important constituent of vegetable ashes, especially of oats and the root-crops.

It is often deficient in the soil, particularly where potatoes have been long cultivated. One of the reasons why *plaster* (sulphate of lime) is so beneficial to the potato crop is undoubtedly that it supplies it with sulphuric acid.

Sulphuric acid is commonly known by the name of *oil vitriol,* and may be purchased for agricultural purposes at a low price. It may be used in a very dilute form (weakened by mixing it with a large quantity of water) to the compost heap, where it will change the ammonia to a sulphate as soon as formed, and thus prevent its loss, as the sulphate of ammonia is not volatile; and, being soluble in water, is useful to plants. Some idea of the value of this compound may be formed from the fact that manufac-

---

What is sulphuric acid commonly called?
How may it be used?
How does it prevent the escape of ammonia?

turers of manures are willing to pay seven cents per lb., or even more, for sulphate of ammonia, to insure the success of their fertilizers. Notwithstanding this, many farmers persist in throwing away hundreds of pounds of *ammonia* every year, as a tax for their ignorance (or indolence), while a small tax in *money*—not more valuable, nor more necessary to their success—for the support of common schools, and the better education of the young, is too often unwillingly paid.

If a tumbler full of sulphuric acid (costing a few cents), be thrown into the tank of the compost heap once a month, the benefit to the manure would be very great.

Where a deficiency of sulphuric acid in the soil is indicated by analysis, it may be supplied in this way, or by the use of plaster or refuse epsom salts.

Care is necessary that *too much* sulphuric acid be not used, as it would prevent the proper decomposition of manures, and would induce a growth of sorrel in the soil by making it *sour*.

In many instances, it will be found profitable to use sulphuric acid in the manufacture of super-phosphate of lime (as directed under the head of 'phosphoric acid,') thus making it perform the double purpose of preparing an available form of phosphate, and of supplying sulphur and sulphuric acid to the plant.

What is the effect of using too much sulphuric acid?

PHOSPHORIC ACID.

We come now to the consideration of one of the most important of all subjects connected with agriculture, that is, *phosphoric acid.*

Phosphoric acid, forming about one half of the ashes of wheat, rye, corn, buck-wheat, and oats; nearly the same proportion of those of barley, peas, beans and linséed; an important ingredient of the ashes of potatoes and turnips; one quarter of the ash of milk and a large proportion of the bones of animals, often exists in the soil in the proportion of only about one or two pounds in a thousand. The cultivation of our whole country has been such, as to take away the phosphoric acid from the soil without returning it, except in very minute quantities. Every hundred bushels of wheat sold contains (and removes permanently from the soil) about *sixty pounds* of phosphoric acid. Other grains, as well as the root crops and grasses, remove likewise a large quantity of it. It has been said by a contemporary writer, that for each cow kept on a pasture through the summer, there is carried off in veal, butter and cheese, not less than *fifty* lbs. of phosphate of lime (bone-earth) on an average.

How large a part of the ashes of grain consists of phosphoric acid?

Of what other substances does it form a leading ingredient?

How many pounds of sulphuric acid are contained in one hundred bushels of wheat?

This would be *one thousand lbs.* for twenty cows; and it shows clearly why old dairy pastures become so exhausted of this substance, that they will no longer produce those nutritious grasses, which are favorable to butter and cheese-making.

That this removal of the most valuable constituent of the soil, has been the cause of more exhaustion of farms, and more emigration, in search of fertile districts, than any other single effect of injudicious farming, is a fact which multiplied instances most clearly prove.

It is stated that the Genesee and Mohawk valleys, which once produced an average of *thirty-five* or *forty bushels* of wheat, per acre, have since been reduced in their average production to *twelve and a half* bushels. Hundreds of similar cases might be stated; and in a large majority of these, could the cause of the impoverishment be ascertained, it would be found to be the removal of the phosphoric acid from the soil.

The evident tendency of cultivation being to continue this murderous system, and to prey upon the vital strength of the country, it is necessary to take such measures as will arrest the outflow of this valuable material. This can never be fully accomplished until laws shall be made preventing the wastes

---

How much phosphate of lime will twenty cows remove from a pasture during a summer?

What has this removal of phosphate of lime occasioned?

How have the Genesee and Mohawk valleys been affected by this removal of phosphoric acid?

of cities and towns. Such laws have existed for a long time in China, and have doubtlessly been the secret of the long subsistence and present prosperity of the millions of people inhabiting that country.

We have, nevertheless, a means of restoring to fertility many of our worn-out lands, and preserving our fertile fields from so rapid impoverishment as they are now suffering. Many suppose that soils which produce good crops, year after year, are inexhaustible, but time will prove to the contrary. They may possess a sufficiently large stock of phosphoric acid, and other constituents of plants, to last a long time, but when that stock becomes so reduced, that there is not enough left for the uses of full crops, the productive power of the soil will yearly decrease, until it becomes worthless. It may last a long time, a century, or even more, but as long as the system is—to *remove every thing, and return nothing,*—the fate of the most fertile soil is evident.

The source mentioned, from which to obtain phosphoric acid, is the bones of animals. These contain large quantities of *phosphate of lime.* They are the receptacles which collect nearly all of the phosphates in crops, which are fed to animals, and are not returned in their excrements. For the grain, etc., sent out of the country, there is no way to be

---

How may this devastation be arrested?
Is any soil inexhaustible?
What is usually the best source from which to obtain phosphoric acid?

repaid except by the importation of this material; but, all that is fed to animals, or to human beings, may, if a proper use be made of their excrement, and of their bones after death, be returned to the soil. With the treatment of animal excrements we are already familiar, and we will now turn our attention to the subject of

BONES.

Bones consist, when dried, of about one third organic matter, and two thirds inorganic matter.

The organic matter consists chiefly of *gelatine*—a compound containing *nitrogen*.

The inorganic part is chiefly *phosphate of lime*.

Hence, we see that bones are excellent, as both organic and mineral manure. The organic part, containing nitrogen, forms *ammonia*, and the inorganic part supplies the much needed *phosphoric acid* to the soil.

Liebig says that, as a producer of ammonia, 100 lbs. of dry bones are equivalent to 250 lbs. of human urine.

Bones are applied to the soil in almost every conceivable form. *Whole bones* are often used in very

---

Of what do dried bones consist?
What is the organic matter of bones?
The inorganic?
What can you say of the use of whole bones?

large quantities ; their action, however, is extremely slow, and it is never advisable to use bones in this form.

Ten bushels of bones, finely ground, will produce larger results, during the current ten years after application, than would ensue from the use of one hundred bushels merely broken, not because the dust contains more fertilizing matter than the whole bones, but because that which it does contain is in a much more available condition. It ferments readily, and produces ammonia, while the ashy parts are exposed to the action of roots.

Bone-black. If bones are burned in retorts, or otherwise protected from the atmosphere, their organic matter will all be driven off, except the carbon, which not being supplied with oxygen cannot escape. In this form bones are called *ivory black*, or *bone-black.* It consists of the inorganic matter, and the carbon of the bones. The nitrogen having been expelled it can make no ammonia, and thus far the original value of bones is reduced by burning ; that is, one ton of bones contains more fertilizing matter before, than after burning ; but one ton of bone black is more valuable than one ton of raw bones, as the carbon is retained in a good form to act as an

---

How does the value of bone dust compare with that of broken bones?

What is the reason of the superiority of bone dust?

How is bone-black made?

Of what does it consist?

absorbent in the soil, while the whole may be crushed or ground much more easily than before being burned. This means of pulverizing bones is adopted by manufacturers, who replace the ammonia in the form of guano, or otherwise ; but it is not to be recommended for the use of farmers, who should not lose the ammonia, forming a part of bones, more than that of other manure.

Composting bones with ashes is a good means of securing their decomposition. They should be placed in a water-tight vessel (such as a cask) ; first, three or four inches of bones, then the same quantity of strong unleached wood ashes, continuing these alternate layers until the cask is full, and keeping them *always wet.* If they become too dry they will throw off an offensive odor, accompanied by the escape of ammonia, and consequent loss of value. In about one year, the whole mass of bones (except, perhaps, those at the top) will be softened, so that they may be easily crushed, and they are in a good condition for manuring. The ashes are, in themselves, valuable, and this compost is excellent for many crops, particularly for Indian corn. A little dilute sulphuric acid, occasionally sprinkled on the upper part of the matter in the cask, will prevent the escape of the ammonia.

Boiling bones under pressure, whereby their gela-

Should farmers burn bones before using them?
How would you compost bones with ashes?
In what way would you prevent the escape of ammonia?

tine is dissolved away, and the inorganic matter left in an available condition, from its softness, is a very good way of rendering them useful; but, as it requires, among other things, a steam boiler, it is hardly probable that it will be largely adopted by farmers of limited means.

Any or all of these methods are good, but bones cannot be used with true economy, except by changing their inorganic matter into

SUPER-PHOSPHATE OF LIME.

Super-phosphate of lime is made by treating phosphate of lime, or the ashes of bones, with *sulphuric acid.*

Phosphate of lime, as it exists in bones, consists of one atom of phosphoric acid and three atoms of lime. It may be represented as

$$\text{Phosphoric acid} \quad \left\{ \begin{array}{l} \text{Lime} \\ \text{Lime} \\ \text{Lime} \end{array} \right.$$

By adding a proper quantity of sulphuric acid with this, it becomes *super*-phosphate of lime; that is, the same amount of phosphoric acid, with a smaller proportion of lime (or a *super*-abundance of

---

What is the effect of boiling bones under pressure?
How is super-phosphate of lime made?
Describe the composition of phosphate of lime, and the chemical changes which take place in altering it to super-phosphate of lime.

phosphoric acid), the sulphuric acid, taking two atoms of lime away from the compound, combined with it making sulphate of lime (plaster). The changes may be thus represented.

Phosphate of lime	Phosphoric acid Lime	Super-phosphate of lime.
	Lime Lime	Sulphate of lime.
Sulphuric acid		

Super-phosphate of lime may be made from whole bones, bone dust, bone-black, or from the pure ashes of bones.

The process of making it from whole bones is slow and troublesome, as it requires a long time for the effect to diffuse itself through the whole mass of a large bone. When it is made in this way, the bones should be *dry*, and the acid should be diluted in many times its bulk of water, and should be applied to the bones (which may be placed in a suitable cask, with a spiggot at the bottom), in quantities sufficient to cover them, about once in ten days; and at the end of that time, one half of the liquid should be drawn off by the spiggot. This liquid is a solution of super-phosphate of lime, containing sulphate of lime, and may be applied to the soil in a liquid form, or through the medium of a compost heap. The object of using so much water is to prevent an incrustation of sulphate of lime on

---

How should sulphuric acid be applied to whole bones?
What is the necessity for so large an amount of water?

the surfaces of the bones, this must be removed by stirring the mass, which allows the next application of acid to act directly on the phosphate remaining. The amount of acid required is about 50 or 60 lbs. to each 100 lbs. of bones. The gelatine will remain after the phosphate is all dissolved, and may be composted with muck, or plowed under the soil, where it will form ammonia.

Bone dust, or *crushed bones*, may be much more easily changed to the desired condition, as the surface exposed is much greater, and the acid can act more generally throughout the whole mass. The amount of acid required is the same as in the other case, but it may be used *stronger*, two or three times its bulk of water being sufficient, if the bones are finely ground or crushed—more or less water should be used according to the fineness of the bones. The time occupied will also be much less, and the result of the operation will be in better condition for manure.

Bones may be made fine enough for this operation, either by grinding, etc., or by boiling under pressure, as previously described ; indeed, by whatever method bones are pulverized, they should always be treated with sulphuric acid before being applied to the soil, as this will more than double their value for immediate use.

Bone-black is chiefly used by manufacturers of

May less water be employed in making super-phosphate from bone dust or crushed bones?

super-phosphate of lime, who treat it with acid the same as has been directed above, only that they grind the black very finely before applying the acid.

Bone ashes, or bones burned to whiteness, may be similarly treated. Indeed, in all of the forms of bones here described, the phosphate of lime remains unaltered, as it is indestructible by heat ; the differences of composition are only in the admixture of organic constituents.

The reason why super-phosphate of lime is so much better than phosphate, may be easily explained. The *phosphate* is very *slowly* soluble in water, and consequently furnishes food to plants slowly. A piece of bone as large as a pea may lie in the soil for years without being all consumed ; consequently, it will be years before its value is returned, and it pays no interest on its cost while lying there. The *super-phosphate* dissolves very *rapidly* and furnishes food for plants with equal facility ; hence its much greater value as a manure.

It is true that the *phosphate* is the most *lasting* manure ; but, once for all, let us caution farmers against considering this a virtue in mineral manures, or in organic manures either, when used on soils con-

---

What other forms of bones may be used in making super-phosphate of lime?

Why is super-phosphate of lime a better fertilizer than phosphate of lime?

What can you say of the *lasting manures?*

taining the proper absorbents of ammonia. They are *lasting*, only in proportion as they are *lazy*. Manures are worthless unless they are in condition to be immediately used. The farmer who wishes his manures to *last* in the soil, and to lose their use, may be justly compared with the *miser*, who buries his gold and silver in the ground for the satisfaction of knowing that he owns it. It is an old and a true saying that "a nimble sixpence is better than a slow shilling."

IMPROVED SUPER-PHOSPHATE OF LIME.

To show the manner in which super-phosphate of lime is perfected, and rendered the best manure for general uses, which has yet been made, containing large quantities of phosphoric acid and a good supply of ammonia,—hereby covering the two leading deficiencies in a majority of soils, it may be well to explain the composition of the *improved super-phosphate of lime* invented by Prof. Mapes.

This manure consists of the following ingredients in the proportions named:—

100 lbs. bone-black (phosphate of lime and carbon).
56 " sulphuric acid.
36 " guano.
20 " sulphate of ammonia.

What are the ingredients of the *improved* super-phosphate of lime?

The sulphuric acid has the before-mentioned effect on the bone-black, and *fixes* the ammonia of the guano by changing it to a sulphate. The twenty pounds of sulphate of ammonia added increase the amount, so as to furnish nitrogen to plants in sufficient quantities to give them energy, and induce them to take up the super-phosphate of lime in the manure more readily than would be done, were there not a sufficient supply of ammonia in the soil.

The addition of the guano, which contains all of the elements of fertility, and many of them in considerable quantities, renders the manure of a more general character, and enables it to produce very large crops of almost any kind, while it assists in fortifying the soil in what is usually its weakest point—phosphoric acid.

Prof. Mapes has more recently invented a new fertilizer called nitrogenized super-phosphate of lime, composed of the improved super-phosphate of lime and blood, dried and ground before mixture, in equal proportions. This manure, from its highly nitrogenous character, theoretically surpasses all others, and probably will be found in practice to have great value ; its cost will be rather greater than guano.

We understand its manufacture will shortly be commenced by a company now forming for that purpose.

Explain the uses of these different constituents.
What is nitrogenized phosphate?

Many farmers will find it expedient to purchase bones, or bone dust, and manufacture their own super-phosphate of lime ; others will prefer to purchase the prepared manure. In doing so, it should be obtained of men of known respectability, as manures are easily adulterated with worthless matters ; and, as their price is so high, that such deception may occasion great loss.

We would not recommend the application of any artificial manure, without first obtaining an analysis of the soil, and knowing *to a certainty* that the manure is needed ; still, when no analysis has been procured, it may be profitable to apply such manures as most generally produce good results—such as stable manure, night soil, the improved super-phosphate of lime ; or, if this cannot be procured, guano.

NEUTRALS.

SILICA.

Silica (or sand) always exists in the soil in sufficient quantities for the supply of food for plants ; but, as has been often stated in the preceding pages, not always in the proper condition. This subject has been so often explained to the student of this book,

---

What should be learned before purchasing amendments for the soil?

What do you know of silica?

that it is only necessary to repeat here, that when the weakness of the straw or stalk of plants grown on any soil indicates an inability in that soil to supply the silicates required for strength, not more sand should be added, but *alkalies*, to combine with the sand already contained in it, and make *soluble silicates* which are available to roots.

Sand is often necessary to stiff clays, as a *mechanical* manure, to loosen their texture and render them easier of cultivation, and more favorable to the distribution of roots, and to the circulation of air and water.

CHLORINE.

Chlorine, a necessary constituent of plants, and often deficient in the soil (as indicated by analysis), may be applied in the form of salt (chloride of sodium), or chloride of lime. The former may be dissolved in the water used to slake lime, and the latter may, with much advantage, be sprinkled around stables and other places where fertilizing gases are escaping, and, after being saturated with ammonia, applied to the soil, thus serving a double purpose.

OXIDE OF IRON.

Nearly all soils contain sufficient quantities of

How may chlorine be applied?

oxide of iron, or iron rust, so that this substance can hardly be required as a manure.

Some soils, however, contain the *proto*xide of iron in such quantities as to be injurious to plants, —see page 86. When this is the case, it is necessary to plow the soil thoroughly, and use such other mechanical means as shall render it open to the admission of air. The *proto*xide of iron will then take up more oxygen, and become the *per*oxide—which is not only inoffensive, but is absolutely necessary to fertility.

OXIDE OF MANGANESE.

This can hardly be called an essential constituent of plants, and is never taken into consideration in manuring lands.

VARIOUS OTHER MINERAL MANURES.

LEACHED ASHES.

Among the mineral manures which have not yet been mentioned—not coming strictly under any of the preceding heads, is the one known as *leached ashes.*

These are not without their benefits, though worth much less than unleached ashes, which, be-

---

How may the protoxide of iron be changed to peroxide?

sides the constituents of those which have been leached, contain much potash, soda, etc.

Farmers' have generally overrated the value of leached ashes, because they contain small quantities of available phosphate of lime, and soluble silicates, in which most old soils are deficient. While we witness the good results ensuing from their application, we should not forget that the fertilizing ingredients of *thirty bushels* of these ashes may be bought in a more convenient form for *ten* or *fifteen cents*, or for less than the cost of spreading the ashes on the soil. In many parts of Long Island farmers pay as much as eight or ten cents per bushel for this manure, and thousands of loads of leached ashes are taken to this locality from the river counties of New York, and even from the State of Maine, and are sold for many times their value, producing an effect which could be as well and much more cheaply obtained by the use of small quantities of super-phosphate of lime and potash.

These ashes often contain a little charcoal (resulting from the imperfect combustion of the wood), which acts as an absorbent of ammonia.

It is sometimes observed that *unleached* ashes,

Why are leached ashes inferior to those that have not been leached?

On what do the benefits of leached ashes depend?

Can these ingredients be more cheaply obtained in another form?

Why do unleached ashes, applied in the spring, sometimes cause grain to lodge?

when applied in the spring, cause grain to lodge. When this is the case, as it seldom is, it may be inferred that the potash which they contain causes so rapid a growth, that the soil is not able to supply silicates as fast as they are required by the plants, but after the first year, the potash will have united with the silica in the soil, and overcome the difficulty.

OLD MORTAR.

Old mortar is a valuable manure, because it contains nitrate of potash and other compounds of nitric acid with alkalies.

These are slowly formed in the mortar by the changing of the nitrogen of the hair (in the mortar) into nitric acid, and the union of this with the small quantities of *potash*, or with the *lime* of the plaster. Nitrogen, presented in other forms, as ammonia, for instance, may be transformed into nitric acid, by uniting with the oxygen of the air, and this nitric acid combines immediately with the alkalies of the mortar.*

The lime contained in the mortar may be useful in the soil for the many purposes accomplished by other lime.

* See Working Farmer, vol. 2, p. 278.

What are the most fertilizing ingredients of old mortar?

GAS HOUSE LIME.

The refuse lime of gas works, where it can be cheaply obtained, may be advantageously used as a manure. It consists, chiefly, of various compounds of sulphur and lime. It should be composted with earth or refuse matter, so as to expose it to the action of air. It should never be used fresh from the gas house. In a few months the sulphur will have united with the oxygen of the air, and become sulphuric acid, which unites with the lime and makes sulphate of lime (plaster), which form it must assume, before it is of much value. Having been used to purify gas made from coal, it contains a small quantity of ammonia, which adds to its value. It is considered a profitable manure in England, at the price there paid for it (forty cents a cartload), and, if of good quality, it may be worth double that sum, especially for soils deficient in plaster, or for such crops as are much benefited by plaster. Its price must, of course, be regulated somewhat by the price of lime, which constitutes a large proportion of its fertilizing parts. The offensive odor of this compound renders it a good protection against many insects.

How may gas-house lime be prepared for use?
Why should it not be used fresh from the gas house?
On what do its fertilizing properties depend?
What use may be made of its offensive odor?

The refuse *liquor of gas works* contains enough ammonia to make it a valuable manure.

SOAPERS' LEY AND BLEACHERS' LEY.

The refuse ley of soap factories and bleaching establishments contains greater or less quantities of soluble silicates and alkalies (especially soda and potash), and is a good addition to the tank of the compost heap, or it may be used directly as a liquid application to the soil. The soapers' ley, especially, will be found a good manure for lands on which grain lodges.

Much of the benefit of this manure arises from the soluble silicates it contains, while its nitrogenous matter,* obtained from those parts of the fatty matters which cannot be converted into soap, and consequently remains in this solution, forms a valuable addition. Heaps of soil saturated with this liquid in autumn, and subjected to the freezings of winter, form an admirable manure for spring use. Mr. Crane, near Newark (N. J.), has long used a mixture of spent ley and stable manure, applied in the fall to trenches plowed in the soil, and has been most successful in obtaining large crops.

* Glycerine, etc.

What use may be made of the refuse léy of soap-makers and bleachers?

What peculiar qualities does soapers' ley possess?

IRRIGATION.

Irrigation does not come strictly under the head of inorganic manures, as it often supplies ammonia to the soil. Its chief value, however, in most cases, must depend on the amount of mineral matter which it furnishes.

The word "irrigation" means simply *watering.* In many districts water is in various ways made to overflow the land, and is removed when necessary for the purposes of cultivation. All river and spring water contains some impurities, many of which are beneficial to vegetation. These are derived from the earth over, or through which, the water has passed, and ammonia absorbed from the atmosphere. When water is made to cover the earth, especially if its rapid motion be arrested, much of this fertilizing matter settles, and is deposited on the soil. The water which sinks into the soil carries its impurities to be retained for the uses of plants. When, by the aid of under-drains, or in open soils, the water passes *through* the soil, its impurities are arrested, and become available in vegetable growth. It is, of course, impossible to say exactly what kind of mineral matter is supplied by water, as that depends on the kind of rock or soil from which the impurities are derived;

On what does the benefit arising from irrigation chiefly depend?
What kind of water is best for irrigation?
How do under-drains increase the benefits of irrigation?

but, whatever it may be, it is generally soluble and ready for immediate use by plants.

Water which has run over the surface of the earth contains both ammonia and mineral matter, while that which has arisen out of the earth, contains usually only mineral matter. The direct use of the water of irrigation as a solvent for the mineral ingredients of the soil, is one of its main benefits.

To describe the many modes of irrigation would be too long a task for our limited space. It may be applied in any way in which it is possible to cover the land with water, at stated times. Care is necessary, however, that it do not wash more fertilizing matter from the soil than it deposits on it, as would often be the case, if a strong current of water were run over it. Brooks may be dammed up, and thus made to cover a large quantity of land. In such a case the rapid current would be destroyed, and the fertilizing matter would settle; but, if the course of the brook were turned, so that it would run in a current over any part of the soil, it might carry away more than it deposited, and thus prove injurious. Small streams turned on to land, from the washing of roads, or from elevated springs, are good means of irrigation, and produce increased fertility, except

---

What is the difference between water which only runs over the surface of the earth, and that which runs out of the earth?

Why should strong currents of water not be allowed to traverse the soil?

where the soil is of such a character as to prevent the water from passing away, in which case it should be under-drained.

Irrigation was one of the oldest means of fertility ever used by man, and still continues in great favor wherever its effects have been witnessed.

MIXING SOILS.

The *mixing of soils* is often all that is necessary to render them fertile, and to improve their *mechanical* condition. For instance, soils deficient in potash, or any other constituent, may have that deficiency supplied, by mixing with them soil containing this constituent in excess.

It is very frequently the case, that such means of improvement are easily availed of. While these chemical effects are being produced, there may be an equal improvement in the mechanical character of the soil. Thus stiff clay soils are rendered lighter, and more easily workable, by an admixture of sand, while light blowy sands are compacted, and made more retentive of manure, by a dressing of clay or of muck.

Of course, this cannot be depended on as a sure means of chemical improvement, unless the soils are previously analyzed, so as to know their require-

---

How are soils improved by mixing?

ments; but, in a majority of cases, the soil will be benefited, by mixing with it soil of a different character. It is not always necessary to go to other locations to procure the soil to be applied, as the sub-soil is often very different from the surface soil, and simple deep plowing will suffice, in such cases, to produce the required admixture, by bringing up the earth from below to mingle it with that of a different character at the surface.

In the foregoing remarks on the subject of mineral manures, the writer has endeavored to point out such a course as would produce the "greatest good to the greatest number," and, consequently, has neglected much which might discourage the farmer with the idea, that the whole system of scientific agriculture is too expensive for his adoption. Still, while he has confined his remarks to the more simple improvements on the present system of management, he would say, briefly, that *no manuring can be strictly economical that is not based on an analysis of the soil, and a knowledge of the best means of overcoming the deficiencies indicated, together with the most scrupulous care of every ounce of evaporating or soluble manure.*

Why may the same effect sometimes be produced by deep plowing?

What is absolutely necessary to economical manuring?

CHAPTER X.

ATMOSPHERIC FERTILIZERS.

It is not common to look on the gases in the atmosphere in the light of manures, but they are decidedly so. Indeed, they are almost the only organic manure ever received by the uncultivated parts of the earth, as well as a large portion of that which is occupied in the production of food for man.

If these gases were not manures ; if there were no means by which they could be used by plants, the fertility of the soil would long since have ceased, and the earth would now be in an unfertile condition. That this must be true, will be proved by a few moments' reflection on the facts stated in the first part of this book. The fertilizing gases in the atmosphere being composed of the constituents of decayed plants and animals, it is as necessary that they should be again returned to the form of organized matter, as it is that constituents taken from the *soil* should not be put out of existence.

AMMONIA.

The *ammonia* in the atmosphere probably cannot be appropriated by the leaves of plants, and

---

Are the gases in the atmosphere manures?
What would be the result if they were not so?

must, therefore, enter the soil to be assimilated by roots. It reaches the soil in two ways. It is either arrested from the air circulating through the soil, or it is absorbed by rains in the atmosphere, and thus carried to the earth, where it is retained by clay and carbon, for the uses of plants. In the soil, ammonia is the most important of all organic manures. In fact, the value of organic manure may be estimated, either by the amount of ammonia which it will yield, or by its power of absorbing ammonia from other sources.

The most important action of ammonia in the soil is the supply of *nitrogen* to plants; but it has other offices which are of consequence. It assists in some of the chemical changes necessary to prepare the matters in the soil for assimilation. Some argue that ammonia *stimulates* the roots of plants, and causes them to take up increased quantities of inorganic matter. The discussion of this question would be out of place here, and we will simply say, that it gives them such vigor that they require increased amounts of ashy matter, and enables them to take this from the soil.

Although, in the course of nature, the atmospheric fertilizers are plentifully supplied to the soil, without the immediate attention of the farmer, it is

---

How is ammonia used by plants?
How may it be carried to the soil?
How may the value of organic manures be estimated?
What effects has ammonia beside supplying food to plants?

not beyond his power to manage them in such a manner as to arrest a greater quantity. The precautions necessary have been repeatedly given in the preceding pages, but it may be well to name them again in this chapter.

The condition of the soil is the main point to be considered. It must be such as to absorb and retain ammonia—to allow water to pass *through* it, and be discharged *below* the point to which the roots of crops are searching for food—and to admit of a free circulation of air.

The power of absorbing and retaining ammonia is not possessed by sand, but it is a prominent property of clay, charcoal, and some other matters named as absorbents. Hence, if the soil consists of nearly pure sand, it will not make use of the ammonia brought to it from the atmosphere, but will allow it to evaporate immediately after a shower. Soils in this condition require additions of absorbent matters, to enable them to use the ammonia received from the atmosphere. Soils already containing a sufficient amount of clay or charcoal, are thus far prepared to receive benefit from this source.

The next point is to cause the water of rains to pass *through* the soil. If it lies on the surface, or

---

To how great a degree can the farmer control atmospheric fertilizers?

What should be the condition of the soil?

What substances are good absorbents in the soil?

How may sandy soils be made retentive of ammonia?

runs off without entering the soil, or even if it only enters to a slight depth, and comes in contact with but a small quantity of the absorbents, it is not probable that the fertilizing matters which it contains will all be abstracted. Some of them will undoubtedly return to the atmosphere on the evaporation of the water; but, if the soil contains a sufficient supply of absorbents, and will allow all rain water to pass through it, the fertilizing gases will all be retained. They will be filtered (or raked) out of the water.

This subject will be more fully treated in Section IV. in connection with under-draining.

Besides the properties just described, the soil must possess the power of admitting a free circulation of air. To effect this, it is necessary that the soil should be well pulverized to a great depth. If, in addition to this, the soil be such as to admit water to pass through, it will allow that circulation of air necessary to the greatest supply of ammonia.

CARBONIC ACID.

Carbonic acid is received from the atmosphere, both by the leaves and roots of plants.

If there is caustic lime in the soil, it unites with it, and makes it milder and finer. It is absorbed by

---

Why does under-draining increase the absorptive power of the soil?

How do plants obtain their carbonic acid?

How does carbonic acid affect caustic lime in the soil?

the water in the soil, and gives it the power of dissolving many more substances than it would do without the carbonic acid. This use is one of very great importance, as it is equivalent to making the minerals themselves more soluble. Water dissolves carbonate of lime, etc., exactly in proportion to the amount of carbonic acid which it contains. We should, therefore, strive to have as much carbonic acid as possible in the water in the soil; and one way, in which to effect this, is to admit to the soil the largest possible quantity of atmospheric air which contains this gas.

The condition of soil necessary for this, is the same as is required for the deposit of ammonia by the same circulation of air.

OXYGEN.

Oxygen, though not taken up by plants in its pure form, may justly be classed among manures, if we consider its effects both chemical and mechanical in the soil.

1. By oxidizing or *rusting* some of the constituents of the soil, it prepares them for the uses of plants.

What power does it give to water?

What condition of the soil is necessary for the reception of the largest quantity of carbonic acid?

May oxygen be considered a manure?

What is the effect of the oxidation of the constituents of the soil?

2. It unites with the *prot*oxide of iron, and changes it to the *per*oxide.

3. If there are *acids* in the soil, which make it sour and unfertile, it may be opened to the circulation of the air, and the oxygen will prepare some of the mineral matters contained in the soil to unite with the acids and neutralize them.

4. Oxygen combines with the carbon of organic matters in the soil, and causes them to decay. The combination produces carbonic acid.

5. It combines with the nitrogen of decaying substances and forms *nitric acid,* which is serviceable as food for plants.

6. It undoubtedly affects in some way the matter which is thrown out from the roots of plants. This, if allowed to accumulate, and remain unchanged, is often very injurious to plants ; but, probably, the oxygen and carbonic acid of the air in the soil change it to a form to be inoffensive, or even make it again useful to the plant.

7. It may also improve the *mechanical* condition of the soil, as it causes its particles to crumble, thus making it finer ; and it roughens the surfaces of particles, making them less easy to move among each other.

How does it affect the protoxide of iron?
How does it neutralize the acids in the soil?
How does it affect its organic parts?
How does it form nitric acid?
How may it affect excrementitious matter of plants?
What effect has it on the mechanical condition of the soil?

These properties of oxygen claim for it a high place among the atmospheric fertilizers.

WATER.

Water may be considered an atmospheric manure, as its chief supply to vegetation is received from the air in the form of rain or dew. Its many effects are already too well known to need farther comment.

The means of supplying water to the soil by the deposit of *dew* will be fully explained in Section IV.

CHAPTER XI.

RECAPITULATION.

MANURES have two distinct classes of action in the soil, namely, *chemical* and *mechanical*.

Chemical manures are those which enter into the construction of plants, or produce such chemical effects on matters in the soil as shall prepare them for use.

Mechanical manures are those which improve

Why may water be considered an atmospheric manure?
What classes of action have manures?
What are chemical manures? Mechanical?

the mechanical condition of the soil, such as loosening stiff clays, compacting light sands, pulverizing large particles, etc.

Manures are of three distinct kinds, namely, *Organic, mineral,* and *atmospheric.*

Organic manures comprise all vegetable and animal matters (except ashes) which are used to fertilize the soil. Vegetable manures supply carbonic acid, and inorganic matter to plants. Animal manures supply the same substances and ammonia.

Mineral manures comprise ashes, salt, phosphate of lime, plaster, etc. They supply plants with inorganic matter. Their usefulness depends on their solubility.

Many of the organic and mineral manures have the power of absorbing ammonia arising from the decomposition of animal manures, as well as that which is brought to the soil by rains—these are called absorbents.

Atmospheric manures consist of ammonia, carbonic acid, oxygen and water. Their greatest usefulness requires the soil to allow the water of rains to pass *through* it, to admit of a free circulation of air among its particles, and to contain a sufficient amount of absorbent matter to arrest and retain all ammonia and carbonic acid presented to it.

What are the three kinds of manures?

What are organic manures, and what are their uses? Mineral Atmospheric?

SECTION FOURTH.

MECHANICAL CULTIVATION.

Manures should never be applied to the soil without regard to its requirements.

Ammonia and carbon are almost always useful, but mineral manures become mere *dirt* when applied to soils not deficient of them.

The only true guide to the exact requirements of the soil is *chemical analysis* ; and this must always be obtained before farming can be carried on with true economy.

Organic manures must be protected against the escape of their ammonia and the leaching out of their soluble parts. One cord of stable manure properly preserved, is worth ten cords which have lost all of their ammonia by evaporation, and their soluble parts by leaching—as is the case with much of the manure kept exposed in open barn-yards.

Atmospheric manures cost nothing, and are of great value when properly employed. In consequence of this, the soil which is enabled to make the largest appropriation of the atmospheric fertilizers, is worth many times as much as that which allows them to escape.

What rule should regulate the application of manures? How must organic manures be managed? Atmospheric?

SECTION FOURTH

MECHANICAL CULTIVATION.

CHAPTER I.

THE MECHANICAL CHARACTER OF SOILS.

THE mechanical character of the soil is well understood from preceding remarks, and the learner knows that there are many offices to be performed by the soil aside from the feeding of plants.

1. It admits the roots of plants, and holds them in their position.

2. By a sponge-like action, it holds water for the uses of the plant.

3. It absorbs moisture from the atmosphere to supply the demands of plants.

What is the first office of the soil?
How does it hold water for the uses of the plant?
How does it obtain a part of its moisture?

4. It absorbs heat from the sun's rays to assist in the process of growth.

5. It admits air to circulate among roots, and supply them with a part of their food, while the oxygen of that air renders available the minerals of the soil; and its carbonic acid, being absorbed by the water in the soil, gives it the power of dissolving, and carrying into roots more inorganic matter than would be contained in purer water.

6. It allows the excrementitious matter thrown out by roots to be carried out of their reach.

All of these actions the soil must be capable of performing, before it can be in its highest state of fertility. There are comparatively few soils now in this condition, but there are also few which could not be profitably rendered so, by a judicious application of the modes of cultivation to be described in the following chapters.

The three great objects to be accomplished are:—

1. To adopt such a system of drainage as will cause all of the water of rains to pass *through* the soil, instead of evaporating from the surface.

2. To pulverize the soil to a considerable depth.

3. To darken its color, and render it capable of absorbing atmospheric fertilizers.

How may it obtain heat?
What is the use of the air circulating among its particles?
Could most soils be brought to the highest state of fertility?
What is the first thing to be done?
Should its color be darkened?

The means used to secure these effects are *under-draining, subsoil and surface-plowing, digging, applying muck, etc.*

CHAPTER II.

UNDER-DRAINING.

THE advantages of *under*-drains over *open* drains are very great.

When open drains are used, much water passes into them immediately from the surface, and carries with it fertilizing parts of the soil, while their beds are often compacted by the running water and the heat of the sun, so that they become water tight, and do not admit water from the lower parts of the soil.

The sides of these drains are often covered with weeds, which spread their seeds throughout the whole field. Open drains are not only a great obstruction to the proper cultivation of the land, but they cause much waste of room, as we can rarely plow nearer than within six or eight feet of them.

There are none of these objections to the use of under-drains, as these are completely covered, and

---

Name some of the means used to secure these effects.
Why are under-drains superior to open drains?

do not at all interfere with the cultivation of the surface.

Under drains may be made with brush, stones, or tiles. Brush is a very poor material, and its use is hardly to be recommended. Small stones are better, and if these be placed in the bottoms of the trenches, to a depth of eight or ten inches, and covered with sods turned upside down, having the earth packed well down on to them, they make very good drains.

TILE DRAINING.

The best under-drains are those made with tiles, or burnt clay pipes. The first form of these used was that called the *horse-shoe tile,* which was in two distinct pieces; this was superseded by a round pipe, and we have now what is called the *sole tile,* which is much better than either of the others.

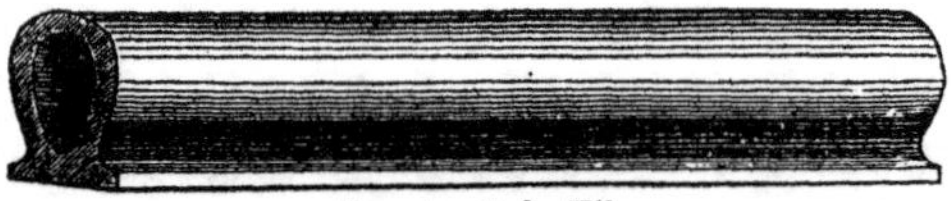

Fig. 4—Sole Tile.

This tile is made (like the horse-shoe and pipe tile) of common brick clay, and is burned the same as bricks. It is about one half or three quarters of an inch thick, and is so porous that water passes di-

---

With what materials may under-drains be constructed?
Describe the tile.

rectly through it. It has a flat bottom on which to stand, and this enables it to retain its position, while making the drain, better than would be done by the round pipe. The orifice through which the water passes is egg-shaped, having its smallest curve at the bottom. This shape is the one most easily kept clear, as any particles of dirt which get into the drain must fall immediately to the point where even the smallest stream of water runs, and are thus removed. An orifice of about two inches is sufficient for the smaller drains, while the main drains require larger tiles.

These tiles are laid, so that their ends will touch each other, on the bottoms of the trenches, and are kept in position by having the earth tightly packed around them. Care must be taken that no space is left between the ends of the tiles, as dirt would be liable to get in and choke the drain. It is advisable to place a sod—grass side down—over each joint, before filling the trench, as this more effectually protects them against the entrance of dirt. There is no danger of keeping the water out by this operation, as it will readily pass through any part of the tiles.

In *digging the trenches* it is not necessary (except in very stony ground) to dig out a place wide enough for a man to stand in, as there are tools made expressly for the purpose, by which a trench may be

---

Why is the sole tile superior to those of previous construction?
How are these tiles laid?
How may the trenches be dug?

dug six or seven inches wide, and to any required depth. One set of these implements consists of a long narrow spade and a hoe to correspond, such as are represented in the accompanying figure.

Fig. 5.

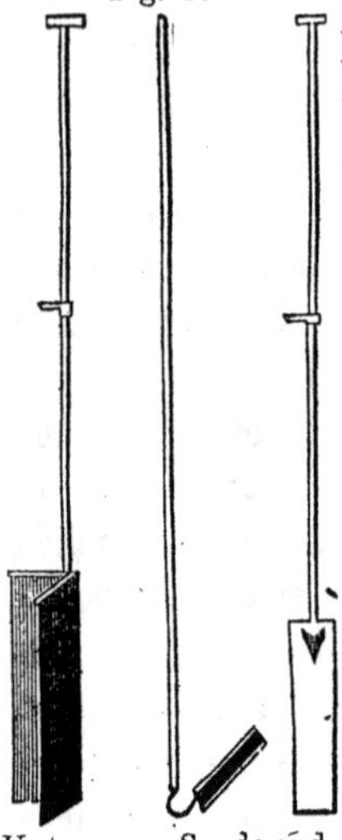

Upton tool. Spade and hoe.

With these tools, and a long light crowbar, for hard soils, trenches may be dug much more cheaply than with the common spade and pickaxe. Where there are large boulders in the soil, these draining tools may dig under them so that they will not have to be removed.

When the trenches are dug to a sufficient depth, the bottoms must be made perfectly smooth, with the required descent (from six inches to a few feet in one hundred feet). Then the tiles may be laid in, so that their ends will correspond, be packed down, and the trenches filled up. Such a drain, if properly constructed, may last for ages. Unlike the stone drain, it is not liable to be frequented by rats, nor choked up by the soil working into it.

The position of the tile may be best represented by a figure, also the mode of constructing stone drains.

Fig. 6.

a—Tile drain trench.
b—Stone drain trench.
c—Sod laid on the stone.

It will be seen that the tile drain is made with much less labor than the stone drain, as it requires less digging, while the breaking up of the stone for the stone drain will be nearly, or quite as expensive as the tiles. Drains made with large stones are not nearly so good as with small ones, because they are more liable to be choked up by animals working in them.*

The *depth* of the drains must depend on the distances at which they are placed. If but *twenty* feet apart, they need be but *three* feet deep; while, if they are *eighty* feet apart, they must be *five* feet deep, to produce the same effect. The reason for this is, that the water in the drained soil is not level, but is higher midway between the drains, than at any other point. It is necessary that this highest point should be sufficiently far from the surface not to interfere with the roots of plants, consequently, as the water line between two drains is *curved*, the

* It is probable that a composition of hydraulic cement and some soluble material will be invented, by which a continuous pipe may be laid in the bottoms of trenches, becoming porous as the soluble material is removed by water.

Why are small stones better than large stones in the construction of drains?

On what must the depth of under-drains depend?

most distant drains must be the deepest. This will be understood by referring to the following diagram.

Fig. 7.

aa—5 feet drains, 80 ft. apart. *bb*—3 feet drains, 20 ft. apart.

The curved line represents the position of the water.

In most soils it will be easier to dig one trench five feet deep, than four trenches three feet deep, and the deep trenches will be equally beneficial; but where the soil is very hard below a depth of three feet, the shallow trenches will be the cheapest, and in such soils they will often be better, as the hard mass might not allow the water to pass down to enter the deeper drains.

By following out these instructions, land may be cheaply, thoroughly, and permanently drained.

Describe the principle which regulates these relative depths and distances. (Black board.)

Which is usually the cheaper plan of constructing drains?

CHAPTER III.

ADVANTAGES OF UNDER-DRAINING.

THE advantages of under-draining are many and important.

1. It entirely prevents drought.

2. It furnishes an increased supply of atmospheric fertilizers.

3. It warms the lower portions of the soil.

4. It hastens the decomposition of roots and other organic matter.

5. It accelerates the disintegration of the mineral matters in the soil.

6. It causes a more even distribution of nutritious matters among those parts of soil traversed by roots.

7. It improves the mechanical texture of the soil.

8. It causes the poisonous excrementitious matter of plants to be carried out of the reach of their roots.

9. It prevents grasses from running out.

10. It enables us to deepen the surface soil.

By removing excess of water—

11. It renders soils earlier in the spring.

12. It prevents the throwing out of grain in winter.

13. It allows us to work sooner after rains.

14. It keeps off the effects of cold weather longer in the fall.

15. It prevents the formation of *acetic* and other organic acids, which induce the growth of sorrel and similar weeds.

16. It hastens the decay of vegetable matter, and the finer comminution of the earthy parts of the soil.

17. It prevents, in a great measure, the evaporation of water, and the consequent abstraction of heat from the soil.

18. It admits fresh quantities of water from rains, etc., which are always more or less imbued with the fertilizing gases of the atmosphere, to be deposited among the absorbent parts of soil, and given up to the necessities of plants.

19. It prevents the formation of so hard a crust on the surface of the soil as is customary on heavy lands.

---

1. Under-draining *prevents drought*, because it gives a better circulation of air in the soil ; (it does so by making it more open). There is always the same amount of water *in* and *about* the surface of the earth. In winter, there is more in the soil than in summer, while in summer, that which has been dried

---

How does under-draining prevent drought?

out of the soil exists in the atmosphere in the form of a *vapor*. It is held in the vapory form by *heat*, which acts as *braces* to keep it distended. When vapor comes in contact with substances sufficiently colder than itself, it gives up its heat—thus losing its braces—contracts, and becomes liquid water.

This may be observed in hundreds of common operations.

It is well known that a cold pitcher in summer robs the vapor in the atmosphere of its heat, and causes it to be deposited on its own surface. It looks as though the pitcher were *sweating*, but the water all comes from the atmosphere, not, of course, through the sides of the pitcher.

If we breathe on a knife-blade, it condenses in the same manner the moisture of the breath, and becomes covered with a film of water.

Stone houses are damp in summer, because the inner surfaces of the walls, being cooler than the atmosphere, cause its moisture to be deposited in the manner described. By leaving a space, however, between the walls and the plaster, this moisture is prevented from being troublesome.

Nearly every night in the summer season, the cold

Why is there less water in the soil in summer than in winter, and where does it exist?

What holds it in its vapory form?

How is it affected by cold substances?

Describe the deposit of moisture on the outside of a pitcher in summer.

What other instances of the same action can be named?

earth receives moisture from the atmosphere in the form of dew.

A cabbage, which at night is very cold, condenses water to the amount of a gill or more.

The same operation takes place in the soil. When the air is allowed to circulate among its lower and *cooler* particles, they receive moisture from the same process of condensation. Therefore, when, by the aid of under-drains, the lower soil becomes sufficiently open to admit of a circulation of air, the deposit of atmospheric moisture will keep the soil supplied with water at a point easily accessible to the roots of plants.

If we wish to satisfy ourselves that this is *practically* correct, we have only to prepare two boxes of finely pulverized soil, one, five or six inches deep, and the other fifteen or twenty inches deep, and place them in the sun at mid-day in summer. The thinner soil will be completely dried, while the deeper one, though it may have been perfectly dry at first, will soon accumulate a large amount of water on those particles which, being lower and more sheltered from the sun's heat than the particles of the thin soil, are made cooler.

With an open condition of subsoil, then, such as may be secured by under-draining, we entirely overcome drought.

How does this principle affect the soil?
Explain the experiment with the two boxes of soil.

2. Under-draining *furnishes an increased supply of atmospheric fertilizers,* because it secures a change of air in the soil. This change is produced whenever the soil becomes filled with water, and then dried; when the air above the earth is in rapid motion, and when the comparative temperature of the upper and lower soils changes. It causes new quantities of the ammonia and carbonic acid which it contains to be presented to the absorbent parts of the soil.

3. Under-draining *warms the lower parts of the soil,* because the deposit of moisture (1) is necessarily accompanied by an abstraction of heat from the atmospheric vapor, and because heat is withdrawn from the whole amount of air circulating through the cooler soil.

When rain falls on the parched surface soil, it robs it of a portion of its heat, which is carried down to equalize the temperature for the whole depth. The heat of the rain-water itself is given up to the soil, leaving the water from one to ten degrees cooler, when it passes out of the drains, than when received by the earth.

There is always a current of air passing from the lower to the upper end of a well constructed drain; and this air is always cooler in warm weather, when it issues from, than when it enters the drain. Its lost heat is imparted to the soil.

How does under-draining supply to the soil an increased amount of atmospheric fertilizers?

How does it warm the lower parts of the soil?

This heating of the lower soil renders it more favorable to vegetation, partially by expanding the spongioles at the end of the roots, thus enabling them to absorb larger quantities of nutritious matters.

4. Under-draining *hastens the decomposition of roots and other organic matters in the soil,* by admitting increased quantities of air, thus supplying *oxygen,* which is as essential in decay as it is in combustion. It also allows the resultant gases of decomposition to pass away, leaving the air around the decaying substances in a condition to continue the process.

This organic decay, besides its other benefits, produces an amount of heat perfectly perceptible to the smaller roots of plants, though not so to us.

5. Draining *accelerates the disintegration of the mineral matters in the soil,* by admitting water and oxygen to keep up the process. This disintegration is necessary to fertility, because the roots of plants can feed only on matters dissolved from *surfaces* ; and the more finely we pulverize the soil, the more surface we expose. For instance, the interior of a stone can furnish no food for plants ; while, if it were finely crushed, it might make a fertile soil.

Any thing, tending to open the soil to exposure, facilitates the disintegration of its particles, and thereby increases its fertility.

How does it hasten the decomposition of roots and other organic matter in the soil?

How does it accelerate the disintegration of its mineral parts?

Why is this disintegration necessary to fertility?

6. Draining *causes a more even distribution of nutritious matters among those parts of soil traversed by roots,* because it increases the ease with which water travels around, descending by its own weight, moving sideways by a desire to find its level, or carried upward by attraction to supply the evaporation at the surface. By this continued motion of the water, soluble matter of one part of the soil may be carried to some other part; and another constituent from this latter position may be carried back to the former. Thus the food of vegetables is continually circulating around among their roots, ready for absorption at any point where it is needed, while the more open character of the soil enables roots to occupy larger portions, making a more even drain on the whole, and preventing the undue impoverishment of any part.

6. Under-drains *improve the mechanical texture of the soil;* because, by the decomposition of its parts, as previously described (4 and 5), it is rendered of a character to be more easily worked; while smooth round particles, which have a tendency to pack, are roughened by the oxidation of their surfaces, and move less easily among each other.

8. Drains *cause the excrementitious matter of*

How does under-draining equalize the distribution of the fertilizing parts of the soil?

Why does this distribution lessen the impoverishment of the soil?

How does under-draining improve the mechanical texture of the soil?

How do drains affect the excrementitious matter of plants?

plants to be carried out of the reach of their roots. Nearly all plants return to the soil those parts of their food, which are not adapted to their necessities, and usually in a form that is poisonous to plants of the same kind. In an open soil, this matter may be carried by rains to a point where roots cannot reach it, and where it may undergo such changes as will fit it to be again taken up.

9. By under-draining, *grasses are prevented from running out*, partly by preventing the accumulation of the poisonous excrementitious matter, and partly because these grasses usually consist of *tillering* plants.

These plants continually reproduce themselves in sprouts from the upper parts of their roots. These sprouts become independent plants, and continue to tiller (thus keeping the land supplied with a full growth), until the roots of the *stools* (or clumps of tillers), come in contact with an uncongenial part of the soil, when the tillering ceases; the stools become extinct on the death of their plants, and the grasses run out.

The open and healthy condition of soil produced by draining prevents the tillering from being stopped, and thus keeps up a full growth of grass until the nutriment of the soil is exhausted.

10. Draining *enables us to deepen the surface-soil,* because the admission of air and the decay of roots

---

Why do they prevent grasses from running out?

render the condition of the subsoil such that it may be brought up and mixed with the surface-soil, without injuring *its quality*.

The second class of advantages of under-draining, arising in the removal of the excess of water in the soil, are quite as important as those just described.

11. *Soils are, thereby, rendered earlier in spring*, because the water, which rendered them cold, heavy, and untillable, is earlier removed, leaving them earlier in a growing condition.

12. *The throwing out of grain in winter* is prevented, because the water falling on the earth is immediately removed instead of remaining to throw up the soil by freezing, as it always does from the upright position taken by the particles of ice.

13. *We are enabled to work sooner after rains*, because the water descends, and is immediately removed instead of lying to be taken off by the slow process of evaporation, and sinking through a heavy soil.

14. *The effects of cold weather are kept off longer in the fall*, because the excess of water is removed, which would produce an unfertile condition on the first appearance of cold weather.

The drains also, from causes already named (3),

How does the removal of water render soils earlier in spring?
Why does it prevent the throwing out of grain in winter?
Why does it enable us to work sooner after rains?
Why does it keep off the effects of cold weather longer in the fall?

keep the soil warmer than before being drained, thus actually lengthening the season, by making the soil warm enough for vegetable growth earlier in spring, and later in autumn.

15. *Lands are prevented from becoming sour by the formation of acetic acid*, etc., because these acids are produced in the soil only when the decomposition of organic matter is arrested by the *antiseptic* (preserving) powers of water. If the water is removed, the decomposition of the organic matter assumes a healthy form, while the acids already produced are neutralized by atmospheric influences, and the soil is restored from sorrel to a condition in which it is fitted for the growth of more valuable plants.

16. The *decay of roots*, etc., is allowed to proceed, because the preservative influence of too much water is removed. Wood, leaves, or other vegetable matter kept continually under water, will last for ages; while, if exposed to the action of the weather, as in under-drained soils, they soon decay.

The presence of too much water, by excluding the oxygen of the air, prevents the *comminution of matters* necessary to fertility.

17. *The evaporation of water, and the consequent abstraction of heat from the soil, is in a great measure prevented* by draining the water out at the *bottom* of

---

How does it prevent lands from becoming sour?

Why does it hasten the decay of roots, and the comminution of mineral matters?

How does it prevent the abstraction of heat from the soil?

the soil, instead of leaving it to be dried off from the surface.

When water assumes the gaseous (or vapory) form, it takes up 1723 times as much *heat* as it contained while a liquid. A large part of this heat is derived from surrounding substances. When water is sprinkled on the floor, it cools the room ; because, as it becomes a vapor, it takes heat from the room. The reason why vapor does not feel hotter than liquid water is, that, while it contains 1723 times as much heat, it is 1723 as large. Hence, a cubic inch of vapor, into which we place the bulb of a thermometer, contains no more heat than a cubic inch of water. The principle is the same in some other cases. A sponge containing a table-spoonful of water is just as *wet* as one twice as large and containing two spoonsful.

If a wet cloth be placed on the head, and the evaporation of its water assisted by fanning, the head becomes cooler—a portion of its heat being taken to sustain the vapory condition of the water.

The same principle holds true with the soil. When the evaporation of water is rapidly going on, by the assistance of the sun, wind, etc., a large quantity of heat is abstracted, and the soil becomes cold.

How much heat does water take up in becoming vapor?
Why does water sprinkled on a floor render it cooler?
Why is not a cubic inch of vapor warmer than a cubic inch of water?
Why does a wet cloth on the head make it cooler when fanned?
How does this principle apply to the soil?

When there is no evaporation taking place, except of water which has been deposited on the lower portions of soil, and carried to the surface by capillary attraction (as is nearly true on under-drained soils), the loss of heat is compensated by that taken from the moisture in the atmosphere by the soil, in the above-named manner.

This cooling of the soil by the evaporation of water, is of very great injury to its powers of producing crops, and the fact that under-drains avoid it, is one of the best arguments in favor of their use. Some idea may, perhaps, be formed of the amount of heat taken from the soil in this way, from the fact that, in midsummer, 25 hogsheads of water may be evaporated from a single acre in twelve hours.

18. When not saturated with water the soil admits the water of rains, etc., which bring with them *fertilizing gases from the atmosphere*, to be deposited among the absorbent parts of soil, and given up to the necessities of the plant. When this rain falls on lands already saturated, it cannot enter the soil, but must run off from the surface, or be removed by evaporation, either of which is injurious. The first, because fertilizing matter is washed away. The second, because the soil is deprived of necessary heat.

19. *The formation of crust on the surface of the soil* is due to the evaporation of water, which is

---

When rains are allowed to *enter* the soil, how do they benefit it?
How do under-drains prevent the formation of a crust on the surface of a soil?

drawn up from below by capillary attraction. It arises from the fact that the water in the soil is saturated with mineral substances, which it leaves at its point of evaporation at the surface. This soluble matter from below, often forms a very hard crust, which is a complete shield to prevent the admission of air with its ameliorating effects, and should, as far as possible, be avoided. Under-draining is the best means of doing this, as it is the best means of lessening the evaporation.

The foregoing are some of the more important reasons why under-draining is always beneficial. Thorough experiments have amply proved the truth of the theory.

The *kinds of soil benefited by under-draining* are nearly as unlimited as the kinds of soil in existence. It is a common opinion, among farmers, that the only soils which require draining are those which are at times covered with water, such as swamps and other low lands; but the facts stated in the early part of this chapter, show us that every kind of soil—wet, dry, compact, or light—receives benefit from the treatment. The fact that land is *too dry*, is as much a reason why it should be drained, as that it is *too wet*, as it overcomes drought as effectually as it removes the injurious effects of too much water.

All soils in which the water of heavy rains does not immediately pass down to a depth of at least

---

What kinds of soil are benefited by under-draining?

thirty inches, should be under-drained, and the operation, if carried on with judgment, would invariably result in profit.

Of the precise *profits* of under-draining this is not the place to speak: many of the agricultural papers contain numerous accounts of its success. It may be well to remark here, that many English farmers give it, as their experience, that under-drains pay for themselves every three years, or that they produce a perpetual profit of 33⅓ per cent., or their original cost. This is not the opinion of *theorists* and *book farmers.* It is the conviction of practical men, who know, *from experience,* that under-drains are beneficial.

The best evidence of the utility of under-draining is the position, with regard to it, which has been taken by the English national government, which affords much protection to the agricultural interests of her people—a protection which in this country is unwisely and unjustly withheld.

In England a very large sum from the public treasury has been appropriated as a fund for loans, on under-drains, which is lent to farmers for the purpose of under-draining their estates, the only security given being the increased value of the soil. The time allowed for payments is twenty years, and only five per cent. interest is charged. By the influence

---

What do English farmers name as the profits of under-draining?
What stand has been taken by the English government with regard to under-draining?

of this patronage, the actual wealth of the kingdom is being rapidly increased, while the farmers themselves, can raise their farms to any desired state of fertility, without immediate investment.

The best proof that the government has not acted injudiciously in this matter is, that private capitalists are fast employing their money in the same manner, and loans on under-drains are considered a very safe investment.

There is no doubt that we may soon have similar facilities for improving our farms, and when we do, we shall find that it is unnecessary to move West to find good soil. The districts nearer market, where the expense of transportation is much less, may, by the aid of under-drains, and a judicious system of cultivation, be made equally fertile.

One very important, though not strictly agricultural, effect of thorough drainage is its removal of certain local diseases, peculiar to the vicinity of marshy or low moist soils. The health-reports in several places in England, show that where *fever and ague* was once common, it has almost entirely disappeared since the general use of under-drains in those localities.

How does under-draining affect the healthfulness of marshy countries?

Describe the sub-soil plow.

CHAPTER IV.

SUB-SOIL PLOWING.

THE *sub-soil plow* is an implement differing in figure from the surface plow. It does not turn a furrow, but merely runs through the sub-soil like a mole—loosening and making it finer by lifting, but allowing it to fall back and occupy its former place. It usually follows the surface plow, entering the soil to the depth of from twelve to eighteen inches below the bottom of the surface furrow.

The best pattern now made (the Mapes plow) is represented in the following figure.

Fig. 8.

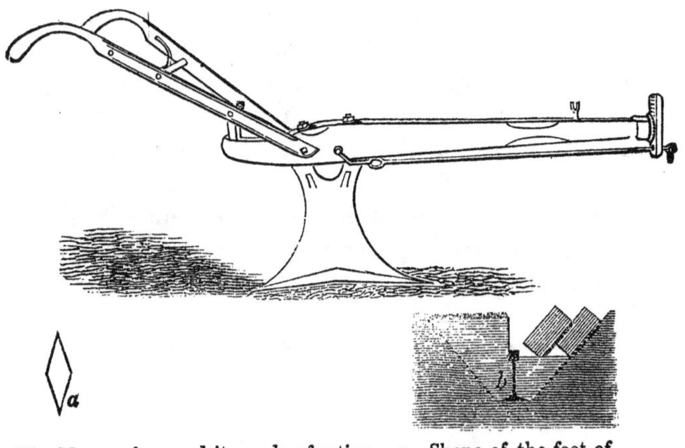

The Mapes plow and its mode of action. *a*—Shape of the foot of the plow. *b*—Its effect on the soil.

The sub-soil plows first made raised the whole soil about eight inches, and required very great power in their use often six, eight, or even ten oxen. The Mapes plow, raising the soil but slightly, may be worked with much less power, and produces equally good results. It may be run to its full depth in most soils by a single yoke of oxen.

Of course a motion in the soil of but one and a half inches is very slight, but it is sufficient to move each particle from the one next to it which, in dry soils, is all that is necessary. Whoever has examined a pile of cannon-balls must have observed that at the points where they touch each other, there is a little rust. In the soil, the same is often the case. Where the particles touch each other, there is such a chemical change produced as renders them fit for the use of plants. While these particles remain in their first position, the changed portions are out of the reach of roots; but, if, by the aid of the sub-soil plow, their position is altered, these parts are exposed for the uses of plants. If we hold in the hand a ball of dry clay, and press it hard enough to produce the least motion among its particles, the whole mass becomes pulverized. On the same principle, the sub-soil plow renders the compact lower soil sufficiently fine for the requirements of fertility.

Describe the Mapes plow.

Why is the motion in the soil of one and a half inches sufficient?

How does the oxydation of the particles of the soil resemble the rusting of cannon balls in a pile?

Notwithstanding its great benefits on land, which is sufficiently dry, sub-soiling cannot be recommended for wet lands ; for, in such case, the rains of a single season would often be sufficient to entirely overcome its effects by packing the subsoil down to its former hardness.

On lands not overcharged with water, it is productive of the best results, it being often sufficient to turn the balance between a gaining and a losing business in farming.

It increases nearly every effect of under-draining ; especially does it overcome drought, by loosening the soil, and admitting air to circulate among the particles of the sub-soil and deposit its moisture on the principle described in the chapter on under-draining.

It deepens the surface-soil, because it admits roots into the sub-soil where they decay and leave carbon, while the circulation of air so affects the mineral parts, that they become of a fertilizing character. The deposit of carbon gives to the subsoil the power of absorbing, and retaining the atmospheric fertilizers, which are more freely presented, owing to the fact that the air is allowed to circulate with greater freedom. As a majority of roots decay in the surface-soil, they there deposit much mineral matter obtained from the sub-soil.

The retention of atmospheric manures is more

Why are the benefits of sub-soiling not permanent on wet lands?
Does sub-soiling overcome drought?
How does it deepen the surface soil?

fully ensured by the better exposure of the clayey portions of the soil.

Those manures which are artificially applied, by being plowed under to greater depths, are less liable to evaporation, as, from the greater amount of soil above them, their escape will more probably be arrested ; and, from the greater prevalence of roots, they are more liable to be taken up by plants.

The sub-soil often contains matters which are deficient in the surface-soil. By the use of the sub-soil plow, they are rendered available.

Sub-soiling is similar to under-draining in continuing the tillering of grasses, and in getting rid of the poisonous excrementitious matter of plants.

When the sub-soil is a thin layer of clay on a sandy bed (as in some plants of Cumberland Co. Maine), the sub-soil plow, by passing through it, opens a passage for water, and often affords a sufficient drainage.

If plants will grow better on a soil six inches deep than on one of three inches, there is no reason why they should not be benefited in proportion, by disturbing the soil to the whole depth to which roots will travel—which is usually more than two feet.

Why is the retention of atmospheric manures ensured by sub-soiling?

Why are organic manures plowed deeply under the soil, less liable to evaporation than when deposited near the surface?

How does sub-soiling resemble under-draining in relation to the tillering of grasses?

When the sub-soil consists of a thin layer of clay on a sandy bed, what use may be made of the sub-soil plow?

The minute rootlets of corn and most other plants, will, if allowed by cultivation, occupy the soil to the depth or thirty-four inches, having a fibre in nearly every cubic inch of the soil for the whole distance. There are very few cultivated plants whose roots would not travel to a depth of thirty inches or more. Even the onion sends its roots to the depth of eighteen inches when the soil is well cultivated.

The object of loosening the soil is to admit roots to a sufficient depth to hold the plant in its position—to obtain the nutriment necessary to its growth—to receive moisture from the lower portions of the soil—and, if it be a bulb, tuber, or tap, to assume the form requisite for its largest development.

It must be evident that roots, penetrating the soil to a depth of two feet, anchor the plant with greater stability than those which are spread more thinly near the surface.

The roots of plants traversing the soil to such great distances, and being located in nearly every part, absorb mineral and other food, in solution in water, only through the *spongioles at their ends.* Consequently, by having these ends in *every part* of the soil, it is *all* brought under contribution, and

---

To how great a depth will the roots of plants usually occupy the soil?

What is the object of loosening the soil?

How are these various effects better produced in deep than in shallow soils?

the amount supplied is greater, while the demand on any particular part may be less than when the whole requirements of plants have to be supplied from a depth of a few inches.

The ability of roots, to assume a natural shape in the soil, and grow to their largest sizes, must depend on the condition of the soil. If it is finely pulverized to the whole depth to which they ought to go, they will be fully developed ; while, if the soil be too hard for penetration, they will be deformed or small. Thus a carrot may grow to the length of two and a half feet, and be of perfect shape, while, if it meet in its course at a depth of eight or ten inches a *cold, hard* sub-soil, its growth must be arrested, or its form injured.

Roots are turned aside by a hard sub-soil, as they would be if received by the surface of a plate of glass.

Add to this the fact that cold, impenetrable sub-soils are *chemically* uncongenial to vegetation, and we have sufficient evidence of the importance, and in many cases the absolute necessity of sub-soiling and under-draining.

It is unnecessary to urge the fact that a garden soil of two feet is more productive than a field soil of six inches ; and it is certain that proper attention to these two modes of cultivation will in a majority of cases make a garden of the field—more than doubling

---

May garden soils be profitably imitated in field culture ?

its value in ease of working, increased produce, certain security against drought, and more even distribution of the demands on the soil—while the outlay will be immediately repaid by an increase of crops.

The sub-soil will be much improved in its character the first year, and a continual advancement renders it in time equal to the original surface-soil, and extending to a depth of two feet or more.

The sub-soil plow is coming rapidly into use. There are now in New Jersey more foundries casting sub-soil plows than there were sub-soil plows in the State six years ago. The implement has there, as well as in many other places, ceased to be a curiosity; and the man who now objects to its use, is classed with him who shells his corn on a shovel over a half-bushel, instead of employing an improved machine, which will enable him to do more in a day than he can do in the "good old way" in a week.

Had we space, we might give many instances of the success of sub-soiling, but the agricultural papers of the present day (at least one of which every farmer should take) have so repeatedly published its advantages, that we will not do so.

In no case will its use be found any thing but satisfactory, except in occasional instances where there is some chemical difficulty in the sub-soil, which an analysis will tell us how to overcome.

Is the use of the sub-soil plow increasing?
Will its use ever injure crops?

As was before stated, its use on wet lands is not advisable until they have been under-drained, as excess of water prevents its effects from being permanent.

CHAPTER V.

PLOWING AND OTHER MODES OF PULVERIZING THE SOIL.

THE advantages of pulverizing the soil, and the *reasons* why it is necessary, are now too well known to need remark. Few farmers, when they plow, dig, or harrow, are enabled to give substantial reasons for so doing. If they will reflect on what has been said in the previous chapters, concerning the supply of mineral food to the plant by the soil, and the effect of air and moisture about roots, they will find more satisfaction in their labor than it can afford when applied without thought.

PLOWING.

The kind of plow used in cultivating the surface-

May the satisfaction attending labor be increased by an understanding of the natural laws which regulate our operations?
On what depends the kind of plow to be used?

soil must be decided by the kind of soil. This question the practical, *observing* farmer will be able to solve.

As a general rule, it may be stated that the plow which runs the *deepest*, with the same amount of force, is the best.

We might enter more fully into this matter but for want of space.

The advantages of *deep plowing* cannot be too strongly urged.

The statement that the *deeper* and the *finer* the soil is rendered, the more productive it will become, is in every respect true, and which no single instance will contradict.

It must not be inferred from this, that we would advise a farmer, who has always plowed his soil to the depth of only six inches, to double the depth at once. Such a practice in some soils would be highly injurious, as it would completely bury the more fertile and better cultivated soil, and bring to the top one which contains no organic matter, and has never been subject to atmospheric influences. This would, perhaps, be so little fitted for vegetation that it would scarcely sustain plants until their roots could reach the more fertile parts below. Such treatment of the soil (turning it upside down) is excellent in *garden* culture, where the great amount of manures

---

What is a general rule with regard to this?
Should deep plowing be immediately adopted? Why?
Why is this course of treatment advisable for garden culture?

applied is sufficient to overcome the temporary barrenness of the soil, but it is not to be recommended for all *field* cultivation, where much less manure is employed.

The course to be pursued in such cases is to *plow one inch deeper each year.* By this means the soil may be gradually deepened to any desired extent. The amount of uncongenial soil which will thus be brought up, is slight, and will not interfere at all with the fertility of the soil, while the elevated portion will become, in one year, so altered by exposure, that it will equal the rest of the soil in fertility.

Often where lime has been used in excess, it has sunk to the sub-soil, where it remains inactive. The slight deepening of the surface plowing would mix this lime with the surface-soil, and render it again useful.

When the soil is light and sandy, resting on a heavy clay sub-soil, or clay on sand, the bringing up of the mass from below will improve the texture of the soil.

As an instance of the success of deep plowing, we call to mind the case of a farmer in New Jersey, who had a field which had yielded about twenty-five bushels of corn per acre. It had been cultivated at ordinary depths. After laying it out in eight step lands (24 feet), he plowed it at all depths from five

---

How should field plowing be conducted?
How does such treatment affect soils previously limed?
How may it sometimes improve sandy or clay soils?

to ten inches, on the different lands, and sowed oats evenly over the whole field. The crop on the five inch soil was very poor, on the six inch rather better, on the seven inch better still, and on the ten inch soil it was as fine as ever grew in New Jersey; it had stiff straw and broad leaves, while the grain was also much better than on the remainder of the field.

There is an old anecdote of a man who died, leaving his sons with the information that he had buried a pot of gold for them, somewhere on the farm. They commenced digging for the gold, and dug over the whole farm to a great depth without finding the gold. The digging, however, so enriched the soil that they were fully compensated for their disappointment, and became wealthy from the increased produce of their farm.

Farmers will find, on experiment, that they have gold buried in their soil, if they will but dig deep enough to obtain it. The law gives a man the ownership of the soil for an indefinite distance from the surface, but few seem to realize that there is *another farm* below the one they are cultivating, which is quite as valuable as the one on the surface, if it were but properly worked.

Fall plowing, especially for heavy lands, is a very good means of securing the action of the frosts of winter to pulverize the soil. If it be a stiff clay, it

---

What kind of soils are benefited by fall plowing?

may be well to throw the soil up into ridges (by ridging and back furrowing), so as to expose the largest possible amount of surface to the freezing and thawing of winter. Sandy soils should not be plowed in the fall, as it renders them too light.

DIGGING MACHINES.

A recent invention has been made in England, known as the digging machine or rotary spade, which —although from having too much gearing between the power and the part performing the labor, it is not adapted to general use—has given such promise of future success, that Mr. Mechi (an agricultural writer of the highest standing) has said that "the plow is doomed." This can hardly be true, for the varied uses to which it may be applied, will guarantee its continuance in the favor of the farmer.

Already, in this country, Messrs. Gibbs & Mapes, have invented a digging machine of very simple construction, which seems calculated to serve an excellent purpose, even in the hands of the farmer of limited means.

Its friends assert that, with one pair of oxen, it will dig perfectly three feet wide, and for a depth of fifteen inches. An experiment with an unperfected machine, in the presence of the writer, seemed to justify their hopes.

What is the digging machine?

This machine thoroughly pulverizes the soil to a considerable depth, and for smooth land must prove far superior to the plow.

THE HARROW AND CULTIVATOR.

The *harrow*, an implement largely used in all parts of the world, to pulverize the soil, and break clods, has become so firmly rooted in the affections of farmers, that it must be a very long time before they can be convinced that it is not the best implement for the use to which it is devoted. It is true that it pulverizes the soil for a depth of two or three inches, and thus much improves its appearance, benefiting it, without doubt, for the earliest stages of the growth of plants. Its action, however, is very defective, because, from the *wedge* shape of its teeth, it continually acts to *pack* the soil; thus—although favorable for the germination of the seed—it is not calculated to benefit the plant during the later stages of its growth, when the roots require the soil to be pulverized to a considerable depth.

The *cultivator* may be considered an *improved harrow*. The principal difference between them being, that while the teeth of the harrow are pointed at the lower end, those of the cultivator are shaped like a small double plow, being large at the bottom

---

Why is the harrow a defective implement?
Why is the cultivator superior to the harrow?

and growing smaller towards the top. They lift the earth up, instead of pressing it downwards, thus loosening instead of compacting the soil.

Many styles of cultivators are now sold at agricultural warehouses. A very good one, for field use, may be made by substituting the cultivator teeth for the spikes in an old harrow frame.

CHAPTER VI.

ROLLING, MULCHING, WEEDING, ETC.

ROLLING.

Rolling the soil with a large roller, arranged to be drawn by a team, is in many instances a good accessory to cultivation. By its means, the following results are obtained :—

1. The soil at the surface is pulverized without the compacting of the lower parts, the area of contact being large.

2. The stones on the land are pressed down so as to be out of the way of the scythe in mowing.

3. The soil is compacted around seeds after sowing in such a manner as to exclude light and to *touch* them in every part, both of which are essential

Name some of the benefits of rolling?

to their germination and to the healthfulness of the plants.

4. The soil is so compacted at the surface, that it is less frequented by *grubs*, etc., than when it is more loose.

5. When the soil is smoothed in this manner, there is less surface exposed for the evaporation of water with its cooling effect.

6. Light sandy lands, by being rolled in the fall, are rendered more compact, and the loosening effects of frequent freezing and thawing are avoided.

Although productive of these various effects, rolling should be adopted only with much care, and should never be applied to very heavy lands, except in dry weather when lumpy after plowing, as its tendency in such cases would be to render them still more difficult of cultivation. Soils in which air does not circulate freely, are not improved by rolling, as it presses the surface-particles still more closely together, and prevents the free admission of the atmosphere.

If well *under-drained*, a large majority of soils would doubtless be benefited by a judicious use of the roller.*

* Field rollers should be made in sections, for ease of turning.

Under what circumstances should the roller be used?

MULCHING.

Mulching (called Gurneyism in England) consists in covering the soil with salt hay, litter, seaweed, leaves, spent tanbark, chips, or other refuse matter.

Every farmer must have noticed that, if a board or rail, or an old brush-heap be removed in spring from soil where grass is growing, the grass afterwards grows in those places much larger and better than in other parts of the field.

This improvement arises from various causes.

1. The evaporation of water from the soil is prevented during drought by the shade afforded by the mulch ; and it is therefore kept in better condition, as to moisture and temperature, than when evaporation goes on more freely. This condition is well calculated to advance the chemical changes necessary to prepare the matters—both organic and mineral—in the soil for the use of plants.

2. By preventing evaporation, we partially protect the soil from losing ammonia resultant from decaying organic matter.

3. A heavy mulch breaks the force of rains, and prevents them from compacting the soil, as would be the result, were no such precaution taken.

4. Mulching protects the surface-soil from freezing as readily as when exposed, and thus keeps it

---

What is mulching?
What are some of its benefits?

longer open for the admission of air and moisture. When unprotected, the soil early becomes frozen; and all water falling, instead of entering as it should do, passes off on the surface.

5. The throwing out of winter grain is often prevented, because this is due to the freezing of the surface-soil.

6. Mulching prevents the growth of some weeds, because it removes from them the fostering heat of the sun.

Many of the best nursery-men keep the soil about the roots of young trees mulched continually. One of the chief arguments for this treatment is, that it prevents the removal of the moisture from the soil and the consequent loss of heat. Also that it keeps up a full supply of water for the uses of the roots, because it keeps the soil cool, and causes a deposit of dew.

7. It also prevents the "baking" of the soil, or the formation of a crust.

It is to be recommended in nearly all cases to sow oats very thinly over land intended for winter fallow after the removal of crops, as they will grow a little before being killed by the frost, when they will fall down, thus affording a very beneficial mulch to the soil.

When farmers spread manure on their fields in the fall to be plowed under in the spring, they benefit

---

Why does mulching take the place of artificial watering?
Why is the late sowing of oats beneficial?
From what arises the chief benefit of top dressing the soil with manure in autumn?

the land by the mulching more than by the addition of fertilizing matter, because they give it the protecting influence of the straw, etc., while they lose much of the ammonia of their manure by evaporation. The same mulching might be more cheaply done with leaves, or other refuse matter, and the ammonia of the manure made available by composting with absorbents.

It is an old and true saying that "snow is the poor man's manure." The reason why it is so beneficial is, chiefly, that it acts as a most excellent mulch. It contains no more ammonia than rain-water does; and, were it not for the fact that it protects the soil against loss of heat, and produces other benefits of mulching, it would have no more advantageous effect. The severity of winters at the North is partially compensated by the long duration of snow.

It is a well known fact that when there is but little snow in cold countries, wheat is very liable to be *winter killed.* The same protection is afforded by artificial mulching.

This treatment is peculiarly applicable to the cultivation of flowers, both in pots and in beds out of doors. It is almost indispensable to the profitable production of strawberries, and many other garden crops, such as asparagus, rhubarb, etc. Many say that the best treatment for trees is to put stones

---

Why is snow particularly beneficial?

about their roots. This is simply *mulching* them, and might be done more cheaply by the use of leaves, copying the action of nature in forests; * for, unless these stones be removed in spring, they will sink and compact the soil in part during open weather.

WEEDING.

If a farmer were asked—what is the use of *weeds?* he might make out quite a list of their benefits, among which might be some of the following:—

1. They shade tender plants, and in a measure serve as a mulch to the ground.

2. Some weeds, by their offensive odor, drive away many insects.

3. They may serve as a green crop to be plowed into the soil, and increase its organic matter.

4. *They make us stir the soil,* and thus increase its fertility.

Still, while thinking out these excuses for weeds, he would see other and more urgent reasons why they should not be allowed to grow.

1. They occupy the soil to the disadvantage of crops.

* The beneficial effects of mulching is so great as to lead us to the conclusion that it has other means of action than those mentioned in this book. Future experiments may lead to more knowledge on this subject.

What are some of the uses of weeds? Their disadvantages?

2. They exclude light and heat from cultivated plants, and thus interfere with their growth.

3. They take up mineral and other matters from the soil, and hold them during the growing season, thus depriving crops of their use.

It is not necessary to argue the injury done by weeds. Every farmer is well convinced that they should be destroyed, and the best means of accomplishing this are of the greatest importance.

In the first place, we should protect ourselves against their increase. This may be done :—

By decomposing all manures in compost, whereby the seeds contained will be killed by the heat of fermentation ; or, if one bushel of salt be mixed through each cord of compost (as before recommended), it will kill seeds as well as grubs,—

By hoeing, or, otherwise, destroying growing weeds before they mature their seeds, and

By keeping the soil in the best chemical condition.

This last point is one of much importance. It is well known that soils deficient in potash, will naturally produce one kind of plants, while soils deficient in phosphoric acid will produce plants of another species, etc. Many soils produce certain weeds which would not grow on them if they were made chemically perfect, as indicated by analysis. It is also believed that those weeds, which naturally grow on

---

How may we protect ourselves against their increase?

Why is it especially important for this purpose to maintain the balance of the soil?

the most fertile soils, are the ones most easily destroyed. There are exceptions (of which the Thistle is one), but this is given as a general rule.

By careful attention to the foregoing points, weeds may be kept from increasing while those already in the soil may be eradicated in various ways, chiefly by mechanical means, such as hoeing, plowing, etc.*

Prof. Mapes says that six bushels of salt annually sown broadcast over each acre of land, will destroy very many weeds as well as grubs and worms.

The *common hoe* is a very imperfect tool for the purpose of removing weeds, as it prepares a better soil for, and replants in a position to grow, nearly as many weeds as it destroys.

The *scuffle-hoe* (or push-hoe) is much more effective, as, when worked by a man walking backwards, and retiring as he works, it leaves nearly all of the weeds on the surface of the soil to be killed by the sun. When used in this way, the earth is not

* It is possible that the excrementitious matter thrown out by some plants may be sufficiently destructive to other kinds to exterminate them from the soil—thus, farmers in Maine say that a single crop of turnips will entirely rid the soil of *witch grass.* This is, undoubtedly, the effect of the excrementitious matter of the turnips. This subject is one of practical importance, and demands close investigation by farmers, which may lead to its being reduced to a system.

How much salt may be used with advantage?
Why is the scuffle-hoe superior to the common hoe?

trodden on after being hoed—as is the case when the common hoe is employed. This treading, besides compacting the soil, covers the roots of many weeds, and causes them to grow again.

Much of the labor of weeding usually performed by men, might be more cheaply done by horses. There are various implements for this purpose, some of which are coming, in many parts of the country, into very general use.

One of the best of these is the *Langdon Horse Hoe*, which is a shovel-shaped plow, to be run one or two inches deep. It has a wing on each side to prevent the earth from falling on to the plants in the rows.. At the rear, or upper edge, is a kind of rake or comb, which allows the earth to pass through, while the weeds pass over the comb and fall on the surface of the soil, to be killed by the heat of the sun. It is a simple and cheap tool, and will perform the work of twenty men with hoes. The hand hoe will be necessary only in the rows.

CULTIVATOR.

The *cultivator*, which was described in the preceding chapter, and of which there are various patterns in use, is excellent for weeding, and for loosening the soil between the rows of corn, etc. The

---

How may much labor be saved in removing weeds?
What is the Langdon horse-hoe?
Describe the *universal* cultivator?

one called the *universal* cultivator, having its side bars made of iron, curved so that at whatever distance it is placed the teeth will point *straight forward,* is a much better tool than those of the older patterns, which had the teeth so arranged that when set for wide rows, they pointed towards the clevis. It is difficult to keep such a cultivator in its place, while the "*universal*" is as difficult to move out of a straight line.

IMPROVED HORSE-HOE.

The *improved horse-hoe* is a combination of the "Langdon" horse hoe and the cultivator, and is the best implement, for many purposes, that has yet been made.*

Fig. 9.

* The improved horse-hoe is made and sold by Ruggles, Nourse & Mason, of Worcester, Mass., and Quincy Hall, Boston.

What is the improved horse-hoe ?

HARVESTING MACHINES.

Until within a comparatively short period, but little attention has been paid to the production of machines for harvesting the various crops.

During the past few years, however, many valuable inventions have appeared. Among these we notice Ketchum's mower, Hussey's mower and reaper, and Wagener's grain and grass seed harvester. The latter machine gathers only the grain and seeds of the crop, leaving the straw to be plowed under the soil, thus maintaining its supply of soluble silicates, and increasing its amount of organic matter. After taking the seed heads from the standing straw and grasses, it thrashes them, blows out the chaff, separates the different kinds of seeds, and discharges them into bags ready for market. It consists of a car containing the machinery ; to this may be attached any required number of horses. The inventor affirms that it has harvested the grain of two acres in one hour, performing the work with accuracy.*

There is much truth in the following proverbs :

" A garden that is well kept, is kept easily."

" You must conquer weeds, or weeds will conquer you."

* This machine is more fully noticed in the advertising pages.

It is almost impossible to give a *recapitulation* of the matters treated in this section, as it is, itself, but an outline of subjects which might occupy our whole book. The scholar and the farmer should understand every principle which it contains, as well as they understand the multiplication table; and their application will be found, in every instance, to produce the best results.

The two great rules of mechanical cultivation are—

THOROUGH UNDER-DRAINING.

DEEP AND FREQUENT DISTURBANCE OF THE SOIL.

What are the two great rules in mechanical cultivation?

SECTION FIFTH.

ANALYSIS.

SECTION FIFTH.

ANALYSIS.

CHAPTER I.

At the present time, when such marked improvements have been, and are still being made, in the practice of agriculture, the farmer cannot be too strongly advised to procure an analysis of his soil, and for obvious reasons.

It has been sufficiently proved that the plant draws from the soil certain kinds of mineral matter, in certain proportions; also, that if the soil do not contain the constituents required, the plants cannot obtain them, and consequently cannot grow. Furthermore, in proportion to the ability of the soil to supply these materials, in exactly the same propor-

---

Why does true practical economy require that the soil should be analyzed?

tion will it, when under good treatment, produce good and abundant crops.

All admit the value and the necessity of manures; they are required to make up deficiencies in the soil, and consequently, they must supply to it the matters which are wanting. In order to know what is wanting, we must know the composition of the soil. This can be learned only by accurate chemical analysis. Such an analysis every farmer must possess before he can conduct his operations with *true practical economy.*

An important question now arises as to whether each farmer can make his own analyses. He cannot do so without long study and practice. The late Prof. Norton said that, at least *two years'* time would be necessary to enable a man to become competent to make a reliable analysis. When we reflect that a farmer may never need more than five or six analyses, we shall see that the time necessary to learn the art would be much more valuable than the cost of the analyses (at $5 or $10 each), setting aside the cost of apparatus, and the fact that while practising in the laboratory, he must not use his hands for any labor that would unfit them for the most delicate manipulations.

Neither will *travelling* chemists be able to make analyses as accurately and as cheaply as those who

---

Can each farmer make his own analyses?
Why will not travelling chemists answer the purpose?
How must an analysis be used?

work in their own laboratories, where their apparatus is not liable to the many injuries consequent on frequent removal. The cost of sending one hundred samples of soil to a distant chemist, would be much less than the expense of having his apparatus brought to the town where his services are required.

The way in which an analysis should be used is a matter of much importance. To a man who knows nothing of chemistry (be he ever so successful a farmer), an analysis, as received from a chemist, would be as useless and unintelligible as though it were written in Chinese ; while, if a chemist who knew nothing of farming, were to give him advice concerning the application of manures, he would be led equally astray, and his course would be any thing but *practical*. It is necessary that chemical and practical knowledge should be combined, and then the value of analysis will be fully demonstrated. The *amount* of knowledge required is not great, but it must be *thorough*. The information contained in this little book is sufficient, but it would be folly for a man to attempt to use an analysis from reading it once hurriedly over. It must be studied and thought on with great care, before it can be of material assistance. The evenings of one winter, devoted to this subject, will enable a farmer to understand the application of analysis to practical farming, especially if other and

---

How may a farmer obtain the requisite knowledge?
When are the services of a consulting agriculturist required?

more compendious works are also read. A less time could hardly be recommended.

Where this attention cannot be given to the subject, the services of a Consulting Agriculturist should be employed to advise the treatment necessary to render fertile the soil analyzed.

Every farmer, however, should learn enough of the principles of agriculture to be able to use an analysis, when procured, without such assistance.*

Nearly all scientific men (all of the highest merit) are unanimous in their conviction of the *practical* value of an analysis of soils; and a volume of instances of their success, with hardly a single failure, might be published.

Prof. Mapes says, in the *Working Farmer*, that he has given advice on hundreds of different soils, and *not a single instance* can be found where he has failed to produce a profit greater than the cost of analysis and advice. Dr. T. C. Jackson, of Boston, the late Prof. Norton, of Yale College, and others, have had universal success in this matter.

Analysis must be considered the only sure road to economical farming.

To select samples of soil for analysis, take a spadeful from various parts of the field—going to exactly the depth to which it has been plowed—until, say a wheel-barrow full, has been obtained. Mix

* See Author's card in the front of the book.

Is there any doubt as to the practical value of analysis?
How should samples of soil for analysis be selected?

this well together, and send about a quart or a pint of it (free from stones) to the chemist. This will represent all of that part of the farm which has been subject to the same cultivation, and is of the same mechanical character. If there are marked differences in the kinds of soil, separate analyses will be necessary.

When an analysis is obtained, a regular debtor and creditor account may be kept with the soil ; and the farmer may know by the composition of the ashes of his crops, and the manures supplied, whether he is maintaining the fertility of his soil.

Prof. Mapes once purchased some land which could not produce corn at all, and by applying only such manures as analysis indicated to be necessary, at a cost of less than $2 per acre, he obtained the first year over *fifty bushels of shelled corn per acre.* The land has since continued to improve, and is as fertile as any in the State. It has produced in one season a sufficient crop of cabbages to pay the expense of cultivation, and over $250 per acre besides, though it was apparently *worthless* when he purchased it.

These are strong facts, and should arouse the farmers of the whole country to their true interests. Let them not call the teachings of science "book-farming," but "prove all things—hold fast that which is good."

Give an instance of the success of treatment according to analysis ?

CHAPTER II.

TABLES OF ANALYSIS.

ANALYSES OF THE ASHES OF CROPS.

No. I.

	Wheat.	Wheat Straw.	Rye.	Rye Straw.
Ashes in 1000 dry parts.....	20	60	24	40
Silica (*sand*)	16	654	5	645
Lime	28	67	50	91
Magnesia	120	33	104	24
Peroxide of Iron...........	7	13	14	14
Potash	237	124	221	174
Soda	91	2	116	3
Chlorine..................		11		5
Sulphuric Acid	3	58	10	8
Phosphoric Acid...........	498	31	496	38

No. II.

	Corn.	Corn Stalks.	Barley.	Barl'y Straw.
Ashes in 1000 dry parts	15	44	28	61
Silica (*sand*)..............	15	270	271	706
Lime	15	86	26	95
Magnesia	162	66	75	32
Peroxide of Iron	3	8	15	7
Oxide of Manganese........				1
Potash	261	96	136	62
Soda	63	277	81	6
Chlorine..................	2	20	1	10
Sulphuric Acid	23	5	1	16
Phosphoric Acid...........	449	171	389	31

No. III.

	Oats.	Oat Straw.	Buck Wheat.	Potatoes.
Ashes in 1000 dry parts	20	51	21	90
Silica (*sand*)...............	7	484	7	42
Lime......................	60	81	67	21
Magnesia	99	38	104	53
Peroxide of Iron...........	4	18	11	5
Potash	{262}	191	87	557
Soda		97	201	19
Chlorine..................	3	32		43
Sulphuric Acid	104	33	22	137
Phosphoric Acid...........	438	27	500	126
Organic Matter............				750 Water.

No. IV.

	Peas.	Beans.	Turnips.	Turnip Tops.
Ashes in 1000 dry parts.....	25	27	76	170
Silica (*sand*)...............	5	12	71	8
Lime......................	53	58	128	233
Magnesia	85	80	48	31
Peroxide of Iron...........	10	6	9	8
Potash	361	336	398	286
Soda	91	106	108	54
Chlorine..................	23	7	37	160
Sulphuric Acid	44	10	131	125
Phosphoric Acid...........	333	378	67	93
Organic Matter............			870 Water.	

No. V.

	Flax.	Linseed.	Meadow Hay.	Red Clover.
Ashes in 1000 dry parts	50	46	60	75
Silica (*sand*)	257	75	344	48
Alumina (*clay*)............	37?			
Lime......................	148	83	196	371
Magnesia	44	146	78	46
Peroxide of Iron..........	36?	9	7	2
Potash	117	240	236	267
Soda	118	45	19	71
Chlorine..................	29	2	28	48
Sulphuric Acid............	32	23	29	60
Phosphoric Acid...........	130	365	58	88

No. VI.

Amount of Inorganic Matter removed from the soil by ten bushels of grains, etc., and by the straw, etc., required in their production—estimated in pounds:

	Wheat.	1200 lbs. Wheat Straw.	Rye.	1620 lbs Rye Straw.
Potash	2.86	8.97	2.51	11.34
Soda	1.04	.12	1.33	.20
Lime......................	.34	4.84	.56	5.91
Magnesia	1.46	2.76	1.18	1.58
Oxide of Iron.............	.08	.94}	.15	.88
Sulphuric Acid............	.03	4.20	.11	.05
Phosphoric Acid...........	6.01	2.22	5.64	2.49
Chlorine..................		.79		.30
Silica	.14	47.16	.05	42.25
Pounds carried off.........	12	72	11½	66

No. VII.

	Corn.	1620 lbs. Corn Stalks.	Oats.	700 lbs. Oat Straw.
Potash	2.78	6.84	1.69	12.08
Soda		19.83		
Lime	.12	6.02	.39	8.39
Magnesia	1.52	4.74	.64	1.59
Oxide of Iron		.57	.02	.78
Sulphuric Acid		.36	.66	1.41
Phosphoric Acid	4.52	12.15	2.80	1.07
Chlorine		1.33	.02	1.36
Silica	.06	19.16	.18	20.32
Pounds carried off	9	71	6½	42

No. VIII.

	Buck Wheat.	Barley.	660 lbs. Barley Straw.	2000 lbs Flax.
Potash	1.01	1.90	2.57	11.78
Soda	2.13	1.18	.23	11.82
Lime	.78	.96	3.88	11.85
Magnesia	1.20	1.00	1.31	9.38
Oxide of Iron	.14	.20	.90	7.32
Sulphuric Acid	.25	.01	.66	3.19
Phosphoric Acid	5.40	5.35	1.25	13.05
Chlorine		.01	.40	2.90
Silica	.09	3.90	28.80	25.71
Pounds carried off	11	14	40	100

No. IX.

	Beans.	1120 lbs. Bean Straw.	Field Peas.	1366 lbs Pea Straw.
Potash	5.54	36.28	5.90	3.76
Soda	1.83	1.09	1.40	
Lime	98.98	13.60	.81	43.93
Magnesia	.28	4.55	1.30	5.50
Oxide of Iron	.10	.20	.15	1.40
Sulphuric Acid	.16	.64	.64	5.43
Phosphoric Acid	7.80	5.00	5.50	3.86
Chlorine	.13	1.74	.23	.08
Silica	.18	4.90	.7	16.02
Pounds carried off	17	68	16	80

No. X.

	1 Ton Turnips.	635 lbs. Turnip Tops.	1 Ton Potatoes.	2000 lbs Red Clover.
Potash	7.14	4.34	27.82	31.41
Soda	.86	.84	.93	8.34
Lime	2.31	3.61	1.03	43.77
Magnesia	.91	.48	2.63	5.25
Oxide of Iron	.23	.13	.26	.23
Sulphuric Acid	2.30	1.81	6.81	7.05
Phosphoric Acid	1.29	1.31	6.25	10.28
Chlorine	.61	2.35	2.13	5.86
Silica	1.36	.13	2.14	5.81
Pounds carried off	17	15	50	118

No. XI.

	2000 lbs. Meadow Hay.	2000 lbs. Cabbage Water 9-10
Potash	18.11	5.25
Soda	1.35	0.20
Lime	22.95	9.45
Magnesia	6.75	2.70
Oxide of Iron	1.69	.25
Sulphuric Acid	2.70	9.60
Phosphoric Acid	5.97	5.60
Chlorine	2.59	2.60
Silica	37.89	.35
Pounds carried off	100	45

No. XII.

Composition of Ashes, leached and unleached, showing their manurial value:

	Oak unleached.	Oak leached.	Beech unleached.	Beech leached.
Potash	84	....	158	....
Soda	56	....	29	....
Lime	750	548	634	426
Magnesia	45	6	113	70
Oxide of Iron	6	....	8	15
Sulphuric Acid	12	....	14	....
Phosphoric Acid	35	8	31	57
Chlorine			2	

No. XIII.

	Birch leached.	Seaweed unleached.	Bituminous Coal unleached.
Potash	....	180	2
Soda	....	210	2
Lime	522	94	21
Magnesia	30	99	2
Oxide of Iron	5	3	40
Sulphuric Acid	....	248	9
Phosphoric Acid	43	52	2
Chlorine	....	98	1

No. XIV.

TOBACCO.

Analysis of the ash of the PLANT [Will & Fresedius]—

Potash	19.55
Soda	0.27
Magnesia	11.07
Lime	48.68
Phosphoric Acid	3.66
Sulphuric Acid	3.29
Oxide of Iron	2.99
Chloride of Sodium	3.54
Loss	6.95
	100.00

Analysis of the ash of the ROOT [Berthier]—

Soluble Matter	12.3
Insoluble Matter	87.7

The Soluble parts consist of nearly—

Carbonic Acid	10.0
Sulphuric Acid	10.3
Muriatic Acid (Chlorine, &c.)	18.26
Potash and Soda	61.44
	100.00

No. XV.

Composition of some of the more common Compounds of Acids and Alkalies.

100 Parts of	Contain of the Alkalies		Contain of the Acids	
Carbonate of Potash (Pearlash)	Potash	68·09	Carbonic	31·91
Bi-Carbonate of Potash (Saleratus)	do.	51·62	Carbonic	48·38
Nitrate of Potash (Saltpetre)	do.	46·56	Nitric	53·44
Silicate of Potash	do.	50·54	Silicic	49·46
Carbonate of Soda	Soda	58·58	Carbonic	41·42
Bi-Carbonate of Soda (Common Soda)*	do.	41·42	Carbonic	58·58
Nitrate of Soda	do.	36·60	Nitric	63·40
Sulphate of Soda (Glauber Salts)*	do.	19·38	Sulphuric	24·85
Silicate of Soda	do.	40·37	Silicic	59·63
Carbonate of Lime (Limestone)	Lime	56·29	Carbonic	43·71
Sulphate of Lime (Plaster Paris)*	do.	32·90	Sulphuric	46·31
Sulphate of Lime (Burned)	do.	41·53	Sulphuric	58·47
Phosphate of Lime	do.	54·48	Phosphoric	45·52
Super-Phosphate of Lime	do.	28·52	Phosphoric	71·48
Silicate of Lime	do.	38·15	Silicic	61·85
Carbonate of Magnesia	Magnesia	48·31	Carbonic	51·69
Sulphate of Magnesia (Epsom Salts)*	do.	16·70	Sulphuric	32·40
Silicate of Alumina	Alumina	27·05	Silicic	72·95
Sulphate of Iron (Green Vitriol)*	Oxide of Iron	27·19	Sulphuric	31·03

* Contain a large amount of Water.

Proximate Analyses of Crops, showing the amount of the different Organic Compounds contained in Grain, Roots, Hay, etc.—estimated in pounds:

		Water.	Husk or Woody Fibre.	Starch, Gum and Sugar.	Gluten, Albumen, Legumin.	Fatty Matter.
10 Bushels.						
Wheat	600 lbs.	90	90	330	87	18
Barley	515 lbs.	77	77	309	70	13
Oats	425 lbs.	68	85	255	70	25
Rye	520 lbs.	62	78	312	65	18
Indian Corn	600 lbs.	84	36	420	72	42
Buck Wheat	425 lbs.	64	106	212	34	2?
Beans	640 lbs.	90	61	256	166	16
Peas	640 lbs.	90	58	320	154	14
2000 lbs.						
Potatoes		1500	80	360	40	6
Turnips		1760	40	180*	30	6
Carrots		1700	60	200*	30	8
Mangold Wurtzel		1700	40	220*	40	?
Meadow Hay		280	600	800	140	70
Clover Hay		280	500	800	186	80
Pea Straw		250	500	900	246	30
Rye Straw		270	900	760	26	?
Corn Stalks		240	500	1040	60	34
100 lbs. Fine Wheat Flour		10		79	11	
100 lbs. Wheat Bran		13		55	19	5

* Pectic Acid.

No. XVII.

Amount of Ash left after burning 1000 lbs. of various plants, ordinarily dry—

Wheat	20		its straw	50
Barley	30		"	50
Oats	40		"	60
Rye	20		"	40
Indian Corn	15		"	50
Pea	30		"	50
Bean	30			
Meadow Hay	50	to 100		
Clover "	90			
Rye Grass "	95			
Potato	8	to 15		
Turnip	5	to 8		
Carrot	15	to 20		

No. XVIII.

MANURES.

HORSE MANURE.

Solid Dung—

Combustible Matter	19.68
Ash	3.07
Water	77.25
	10,000

Composition of the Ash—

Silica	62.40
Potash	11.30
Soda	1.98
Oxide of Iron	1.17
Lime	4.63
Magnesia	3.84
Oxide of Manganese	2.13
Phosphoric Acid	10.49
Sulphuric Acid	1.89
Chlorine	0.03
Loss	0.14
	100.00

No. XIX.

NIGHT SOIL.

Solid (Ash)—

Earthy Phosphates and a trace of Sulphate of Lime	100
Sulphate of Soda and Potash, and Phosphate of Soda	8
Carbonate of Soda	8
Silica	16
Charcoal and Loss	18
	150

Urine

Urea*	30.10	
Uric Acid	1.00	
Sal Ammoniac*	1.50	
Lactic Acid, etc.	17.14	
Mucus	.32	
Sulphate of Potash	3.71	
Sulphate of Soda	3.16	
Phosphate of Ammonia*	1.65	
Earthy Phosphates	3.94	
Salt (Chloride of Sodium)	4.45	
Silica	0.03	
		67.00
Water		933.00
		1000.00

* Supply Ammonia.

No. XX.

COW MANURE.

Solid (Ash)—

Phosphates	20.9
Peroxide of Iron	8.8
Lime	1.5
Sulphate of Lime (Plaster)	3.1
Chloride of Potassium	trace
Silica	63.7
Loss	2.0
	100.0

No. XXI.

COMPARATIVE VALUE OF THE URINE OF DIFFERENT ANIMALS.

	Solid Matter. Organic.	Inorganic.	Total.
Man	23.4	7.6	31
Horse	27.	33.	60
Cow	50.	20.	70
Pig	56.	18.	74
Sheep	28.	12.	40

No. XXII.

GUANO.

Water	6.40
Ammonia	2.71
Uric Acid	34.70
Oxalic Acid, etc.	26.79
Fixed Alkaline Salts.	
Sulphate of Soda	2.94
Phosphate of Soda	.48
Chloride of Sodium (salt)	.86
Earthy Salts.	
Carbonate of Lime	1.36
Phosphates	19.24
Foreign Matter.	
Silicious grit and sand	4.52
	100.00

For the analysis of fertile and barren soils, see page 72.

THE PRACTICAL FARMER.

THE PRACTICAL FARMER.

WHO is the *practical farmer?* Let us look at twc pictures and decide.

Here is a farm of 100 acres in ordinary condition. It is owned and tilled by a hard-working man, who, in the busy season, employs one or two assistants. The farm is free from debt, but it does not produce an abundant income ; therefore, its owner cannot afford to purchase the best implements, or make other needed improvements ; besides, he don't *believe* in such things. His father was a good solid farmer ; so was his grandfather ; and so is he, or thinks he is. He is satisfied that 'the good old way' is best, and he sticks to it. He works from morning till night ; from spring till fall. In the winter, he *rests,* as much as his lessened duties will allow. During this time, he reads little, or nothing. Least of all does he read about farming. He don't want to learn how to dig potatoes out of a book. Book farming is nonsense. Many other similar ideas keep him from agricultural reading. His house is comfortable, and his barns are quite as good as his neighbors', while his farm gives him a living. It is true that his soil does not produce as much as it did ten years ago; but prices are better, and he is satisfied.

Let us look at his premises, and see how his affairs are managed. First, examine the land. Well, it is good fair land. Some of it is a little springy, but is not to be called *wet.* It will produce a ton and a half of hay to the acre—it used to produce two tons. There are some stones on the land, but not enough in his estimation to do harm. The plowed fields are pretty good; they will produce 35 bushels of corn, 13 bushels of wheat, or 30 bushels of oats per acre, when the season is not dry. His father used to get more; but, somehow, the *weather* is not so favorable as it was in old times. He has thought of raising root crops, but they take more labor than he can afford to hire. Over, in the back part of the land there is a muck-hole, which is the only piece of *worthless* land on the whole farm.

Now, let us look at the barns and barn-yards. The stables are pretty good. There are some wide cracks in the siding, but they help to ventilate, and make it healthier for the cattle. The manure is thrown out of the back windows, and is left in piles under the eaves on the sunny side of the barn. The rain and sun make it nicer to handle. The cattle have to go some distance for water; and this gives them exercise. All of the cattle are not kept in the stable; the fattening stock are kept in the various fields, where hay is fed out to them from the stack. The barn-yard is often occupied by cattle, and is covered with their manure, which lies there until it is carted on to the land. In the shed are the tools of the farm, consisting of carts, plows—not deep plows, this farmer thinks it best to have roots near the surface of the soil where they can have the benefit of the sun's heat, a harrow, hoes, rakes, etc. These tools are all in good order; and, unlike those of his less prudent neighbor, they are protected from the weather.

The crops are cultivated with the plow and hoe, as they have been since the land was cleared, and as they always will be until this man dies.

Here is the 'practical farmer' of the present day. Hard working, out of debt, and economical—of dollars and cents, if not of soil and manures. He is a better farmer than two thirds of the three millions of farmers in the country. He is one of the best farmers in his town—there are but few better in the county, not many in the State. He represents the better class of his profession.

With all this, he is, in matters relating to his business, an unreading, unthinking man. He knows nothing of the first principles of farming, and is successful by the *indulgence* of nature, not because he understands her, and is able to make the most of her assistance.

This is an unpleasant fact, but it is one which cannot be denied. We do not say this to disparage the farmer, but to arouse him to a realization of his position and of his power to improve it.

But let us see where he is wrong.

He is wrong in thinking that his land does not need draining. He is wrong in being satisfied with one and a half tons of hay to the acre when he might easily get two and a half. He is wrong in not removing as far as possible every stone that can interfere with the deep and thorough cultivation of his soil. He is wrong in reaping less than his father did, when he should get more. He is wrong in ascribing to the weather, and similar causes, what is due to the actual impoverishment of his soil. He is wrong in not raising turnips, carrots, and other roots, which his winter stock so much need, when they might be raised at a cost of less than one third of their value as food. He is wrong in considering worthless a deposit of muck, which is a mine of wealth if properly employed. He is wrong in *ventilating* his stables at the cost of *heat*. He is wrong in his treatment of his manures, for he loses more than one half of their value from evaporation, fermentation, and leaching. He is wrong in not having water at hand for his cattle—their exercise detracts from their accumulation of fat and their production of heat, and it exposes them to cold. He is wrong in not protecting his fattening stock from the cold of winter; for, under exposure to cold, the food, which would otherwise be used in the formation of *fat*, goes to the production of the animal heat necessary to counteract the chilling influence of the weather, p. 50. He is wrong in allowing his manure to lie unprotected in the barn-yard. He is wrong in not adding to his tools the deep surface plow, the subsoil plow, the cultivator, and many others of improved construction. He is wrong in cultivating with the plow and hoe, those crops which could be better or more cheaply managed with the cultivator or horse-hoe. He is wrong in many things more, as we shall see if we examine all of his yearly routine of work. He is right in a few things ; and but a few, as he himself would admit, had he that knowledge of his business which he could obtain in the leisure hours of a single winter. Still, he thinks himself a *practical* farmer. In twenty years, we shall have fewer such, for our young men have the mental capacity and mental energy necessary to raise them to the highest point of practical education, and to that point they are gradually but surely rising.

Let us now place this same farm in the hands of an educated and understanding cultivator ; and, at the end of five years, look at it again.

He has sold one half of it, and cultivates but fifty acres. The money for which the other fifty were sold has been used in the improvement of the farm. The land has all been under-drained, and shows the many improvements consequent on such treatment. The stones and small rocks have been removed, leaving the surface of the soil smooth, and allowing the use of the sub-soil plow, which with the under-drains have more than doubled the productive power of the farm. Sufficient labor is employed to cultivate with improved tools, extensive root crops, and they invariably give a large yield. The grass land produces a yearly average of 2½ tons of hay per acre. From 80 to 100 bushels of corn, 30 bushels of wheat, and 45 bushels of oats are the average of the crops reaped. The soil has been analyzed, and put in the best possible condition, while it is yearly supplied with manures containing every thing taken away in the abundant crops. The analysis is never lost sight of in the regulation of crops and the application of manures. The *worthless* muck bed was retained, and is made worth one dollar a load to the compost heap, especially as the land requires an increase of organic matter. A new barn has been built large enough to store all of the hay produced on the farm. It has stables, which are tight and warm, and are well ventilated *above* the cattle. The stock being thus protected from the loss of their heat, give more milk, and make more fat on a less amount of food than they did under the old system. Water is near at hand, and the animals are not obliged to over exercise. The manure is carefully composted, either under a shed constructed for the purpose with a tank and pump, or is thrown into the cellar below, where the hogs mix it with a large amount of muck, which has been carted in after being thoroughly decomposed by the lime and salt mixture.

They are thus protected against all loss, and are prepared for the immediate use of crops. No manures are allowed to lie in the barn-yard, but they are all early removed to the compost heap, where they are preserved by being mixed with carbonaceous matter. In the tool shed, we find deep surface-plows, sub-soil plows, cultivators, horse-hoes, seed-drills, and many other valuable improvements.

This farmer takes one or more agricultural papers, from which he learns many new methods of cultivation, while his knowledge of the *reasons* of various agricultural effects enables him to discard the injudicious suggestions of mere *book farmers* and uneducated dreamers.

Here are two specimens of farmers. Neither description is over-drawn. The first is much more careful in his operations than the majority of our rural population. The second is no better than many who may be found in America.

We appeal to the common sense of the reader of this work to know which of the two is the *practical farmer*—let him imitate either as his judgment shall dictate.

FINIS.

EXPLANATION OF TERMS.

ABSORB—to soak in a liquid or a gas.

ABSTRACT—to take from.

ACID—sour; a sour substance.

AGRICULTURE—the art of cultivating the soil.

ALKALI—the direct opposite of an *acid*, with which it has a tendency to unite.

ALUMINA—the base of clay.

ANALYSIS—separating into its primary parts any compound substance.

CARBONATE—a compound, consisting of carbonic acid and an alkali.

CAUSTIC—burning.

CHLORIDE—a compound containing chlorine.

CLEVIS—that part of a plow by which the drawing power is attached.

DECOMPOSE—to separate the constituents of a body from their combinations, forming new kinds of compounds.

DIGESTION—the decomposition of food in the stomach and intestines of animals (agricultural).

DEW—deposit of the insensible vapor of the atmosphere on cold bodies.

EXCREMENT—the matter given out by the organs of plants and animals, being those parts of their food which they are unable to assimilate.

FERMENTATION—a kind of decomposition.

GAS—air—aeriform matter.

GURNEYISM—see *Mulching.*

INGREDIENT—component part.

INORGANIC—mineral, or earthy.

MOULDBOARD—that part of a surface plow which turns the sod.

MULCHING—covering the soil with litter, leaves, or other refuse matter. See p. 247.

NEUTRALIZE—To overcome the characteristic properties of.

ORGANIC MATTER—that kind of matter which at times possesses an organized (or living) form, and at others exists as a gas in the atmosphere.

OXIDE—a compound of oxygen with a metal.

PHOSPHATE—a compound of phosphoric acid with an alkali.

PROXIMATE—an organic compound, such as wood, starch, gum, etc.; a product of life.

PUNGENT—pricking.

PUTREFACTION—rotting.

SATURATE—to *fill* the pores of any substance, as a sponge with water, or charcoal with ammonia.

SILICATE—a compound of silica with an alkali.

SOLUBLE—capable of being dissolved.

SOLUTION—a liquid containing another substance dissolved in it.

SATURATED SOLUTION—one which contains as much of the foreign substance as it is capable of holding.

SPONGIOLES—the mouths at the ends of roots.

SULPHATE—a compound of sulphuric acid with an alkali.

VAPOR—gas.

www.ingramcontent.com/pod-product-compliance
Lightning Source LLC
LaVergne TN
LVHW052254200926
848481LV00002B/360